# NATIONAL RIVERS AUTHORITY
# ANNUAL R&D REVIEW - 1994

**Covering:**

**Work completed in 1993/94**

**Programme for 1994/95**

**Review of R&D from 1989/90 to 1994/95**

**R&D Section**
**Water Management and Science Directorate**
**National Rivers Authority**
**Waterside Drive**
**Aztec West**
**Almondsbury**
**Bristol BS12 4UD**

LONDON: HMSO

# CONTENTS

| | Page |
|---|---|
| GLOSSARY | i |
| LIST OF FIGURES | iii |
| LIST OF TABLES | iii |
| EXECUTIVE SUMMARY | v |
| 1. INTRODUCTION | 1 |
| 2. REVIEW OF R&D IN THE NRA | 7 |
| 3. WATER QUALITY | 14 |
| 4. WATER RESOURCES | 35 |
| 5. FLOOD DEFENCE | 42 |
| 6. FISHERIES | 54 |
| 7. RECREATION AND NAVIGATION | 60 |
| 8. CONSERVATION | 63 |
| 9. CROSS-FUNCTIONAL ISSUES | 69 |
| APPENDIX 1 - R&D Programme for 1994/95 | 77 |
| APPENDIX 2 - R&D Personnel and Management Information | 98 |
| LIST OF R&D OUTPUTS (September 1989 to September 1994) | see inside back cover |

# GLOSSARY

| | | | |
|---|---|---|---|
| ADAS | Agricultural Development and Advisory Service | EPSRC | Engineering and Physical Science Research Council |
| AEA | Atomic Energy Authority | EQS | Environmental Quality Standard |
| ATM | Airborne Thematic Mapper | EU | European Union |
| BBC | British Broadcasting Corporation | FWR | Foundation for Water Research |
| BGS | British Geological Survey | GANDOLF | Generating Advanced Nowcasts for Deployment in Operational Land-Surface Flood Forecasting |
| BSRIA | Building Services Research and Information Association | | |
| | | GPP | Groundwater Protection Policy |
| BW | British Waterways | GQA | General Quality Assessment |
| CASI | Compact Airborne Spectral Imager | GST | Glutathione-S-Transferase |
| CCW | Countryside Council for Wales | HMIP | Her Majesty's Inspectorate of Pollution |
| CEC | Commission for European Communities | HYREX | Hydrological Radar Experiment |
| CEN | Council for European Nations | ICOLE | International Centre of Landscape Ecology (Loughborough University) |
| CES | Consultants in Environmental Science | | |
| CEST | Centre for Exploitation of Science and Technology | IFE | Institute of Freshwater Ecology |
| | | IH | Institute of Hydrology |
| CIRIA | Construction Industry Record and Information Association | ITE | Institute of Terrestrial Ecology |
| CSO | Combined Sewer Overflows | JNCC | Joint Nature Conservation Committee |
| DoE | Department of the Environment | LIFE | Financial Instrument for the Environment |
| DoH | Department of Health | MAFF | Ministry of Agriculture, Fisheries and Food |
| DTA | Direct Toxicity Assessment | Met Office | Meteorological Office |
| DTI | Department of Trade and Industry | NCC | Nature Conservation Council (now English Nature) |
| DTp | Department of Transport | | |
| EA | Environmental Assessment | NERC | Natural Environment Research Council |
| EC | European Community | NFA | National Federation of Anglers |
| ECUS | Environmental Consultancy - University of Sheffield | NOP | National Opinion Poll |
| | | NRA | National Rivers Authority |
| EN | English Nature | OFWAT | Office of Water Services |

| | | | |
|---|---|---|---|
| **OI** | Operational Investigation | **SNH** | Scottish Natural Heritage |
| **PCDD** | Polychlorinated dibenzo-p-dioxins | **SOAFD** | Scottish Office Agriculture and Fisheries Department |
| **PCDF** | Polychlorinated dibenzofurans | **SoS** | Standards of Service |
| **PHABSIM** | Physical Habitat Simulation Model | **SRD** | Safety and Reliability Directorate (AEA) |
| **PML** | Plymouth Marine Laboratory | **SSLRC** | Soil Survey and Land Research Centre |
| **R&D** | Research and Development | **STEP** | Science and Technology for Environmental Protection |
| **RHS** | River Habitat Survey | **SWQOs** | Statutory Water Quality Objectives |
| **RIVPACS** | River Invertebrate Prediction and Classification System | **TBC** | Toxicity-based Consent |
| **RPB** | River Purification Board | **UK** | United Kingdom |
| **RTD** | Research and Technological Development | **UKASTA** | United Kingdom Agricultural Supply Trade Association |
| **SDIA** | Soap and Detergent Industries Association | **UKWIR** | UK Water Industry Research Limited |
| **SERC** | Science and Engineering Research Council (now EPSRC) | **WHO** | World Health Organisation |
| **SETAC** | Society for Environmental Toxicology and Chemistry | **WOAD** | Welsh Office Agriculture Department |
| **SNIFFER** | Scotland and Northern Ireland Forum for Environmental Research | **WRc** | Water Research Centre |
| | | **WSA** | Water Services Association |

## LIST OF FIGURES

| Figure | Title | Page |
|---|---|---|
| 1.1 | Type of R&D in the 1994/95 R&D Programme | 6 |
| 1.2 | Primary purpose of R&D in the 1994/95 R&D Programme | 6 |
| 1.3 | Actual and planned expenditure on R&D in the NRA | 6 |
| 2.1 | The process behind the R&D Programme | 7 |
| 2.2 | Projects developed through the R&D Programme 1989/90 - 1993/94 | 8 |
| 2.3 | Outputs produced through R&D Projects 1989 - 1993 | 13 |
| 3.1 | Actual and planned expenditure on Water Quality R&D up to 1996/97 | 34 |
| 4.1 | Actual and planned expenditure on Water Resources R&D up to 1996/97 | 41 |
| 5.1 | Actual and planned expenditure on Flood Defence R&D up to 1996/97 | 53 |
| 6.1 | Actual and planned expenditure on Fisheries R&D up to 1996/97 | 59 |
| 7.1 | Actual and planned expenditure on Recreation and Navigation R&D up to 1996/97 | 62 |
| 8.1 | Actual and planned expenditure on Conservation R&D up to 1996/97 | 68 |
| 9.1 | Actual and planned expenditure on Cross-Functional Issues R&D up to 1996/97 | 73 |

## LIST OF TABLES

| Table | Title | Page |
|---|---|---|
| 1.1 | NRA Fellowships | 3 |
| 3.1 | Standards and Classification Schemes (Topic A1) R&D Programme (1994/95) | 15 |
| 3.2 | Monitoring - Strategy and Reporting (Topic A2) R&D Programme (1994/95) | 18 |
| 3.3 | Analytical Techniques (Topic A3) R&D Programme (1994/95) | 21 |
| 3.4 | Instrumentation and Field Techniques (Topic A4) R&D Programme (1994/95) | 22 |
| 3.5 | Biological Assessment (Topic A5) R&D Programme (1994/95) | 23 |
| 3.6 | Consenting and Discharge Impact (Topic A6) R&D Programme (1994/95) | 25 |
| 3.7 | Rural Land Use (Topic A7) R&D Programme (1994/95) | 27 |
| 3.8 | Groundwater Pollution (Topic A8) R&D Programme (1994/95) | 31 |
| 3.9 | Pollution Prevention (Topic A9) R&D Programme (1994/95) | 33 |
| 4.1 | Hydrometric Data (Topic B1) R&D Programme (1994/95) | 36 |
| 4.2 | Flow Regimes (Topic B2) R&D Programme (1994/95) | 37 |
| 4.3 | Water Resources Management (Topic B3) R&D Programme (1994/95) | 39 |
| 4.4 | Groundwater Protection (Topic B4) R&D Programme (1994/95) | 40 |
| 5.1 | Fluvial Defences and Processes (Topic C1) R&D Programme (1994/95) | 43 |
| 5.2 | River Flood Forecasting (Topic C2) R&D Programme (1994/95) | 46 |
| 5.3 | Catchment Appraisal and Control (Topic C3) R&D Programme (1994/95) | 47 |
| 5.4 | Operational Management (Topic C4) R&D Programme (1994/95) | 48 |
| 5.5 | Coastal and Tidal Defences and Processes (Topic C6) R&D Programme (1994/95) | 50 |
| 5.6 | Response to Emergencies (Topic C8) R&D Programme (1994/95) | 53 |
| 6.1 | Fisheries Resources (Topic D1) R&D Programme (1994/95) | 55 |
| 6.2 | Environmental and Biological Influences (Topic D2) R&D Programme (1994/95) | 56 |
| 6.3 | Fisheries Management (Topic D3) R&D Programme (1994/95) | 58 |
| 7.1 | Recreation and Navigation (Topic E1) R&D Programme (1994/95) | 61 |
| 8.1 | Conservation Resource Appraisal and Impact Assessment (Topic F1) R&D Programme (1994/95) | 64 |
| 8.2 | Conservation Management (Topic F2) R&D Programme (1994/95) | 66 |
| 9.1 | Cross-Functional Issues (Topic G1) R&D Programme (1994/95) | 70 |

| | | |
|---|---|---|
| A2.1 | Compostion of Research and Development Committee | 99 |
| A2.2 | Commissioners and Topic Leaders | 100 |
| A2.3 | Regional R&D Coordinators | 101 |
| A2.4 | R&D Project Leaders for the 1994/95 Programme | 102 |

# EXECUTIVE SUMMARY

This Annual R&D Review brings together key elements of the National Rivers Authority's Research and Development (R&D) Programme.

The publication of the NRA's R&D Strategy alongside those for its other core functions, in 1993, provided a clear vision of the medium-term water-related R&D requirements. This Review summarises progress against the strategic goals set by the NRA, as well as covering the key achievements in its first five years of operation.

In developing a strategic R&D Programme, the NRA has ensured that projects:

- build on a sound scientific base;

- address clearly identified business requirements;

- produce practical outputs for use by its staff;

- work in partnership with organisations who have a mutual interest in the environment; and

- contribute to a better understanding, management and protection of the environment.

Since 1989, the NRA has commissioned a range of R&D projects covering not only its full range of functions, but also spanning a variety of scientific disciplines. The use of high technology such as airborne remote sensing techniques has provided the NRA with up-to-date operational tools which have enabled it to manage the environment through a better understanding of the spatial pressures placed upon it. In this respect, the NRA is now the largest single user of airborne remote sensing techniques and information in Europe.

The introduction of water quality monitoring instrumentation has also been driven through practical R&D outputs. In 1993, the NRA successfully prosecuted a large chemical company using an automatic monitor and sampler known as CYCLOPS. Versions of this instrument are now in operation in a number of NRA Regions.

As the NRA has taken a strategic view of the management of the water resources in England and Wales, the need for further information on the properties of the major aquifers has become clear. Collaborative work with the British Geological Survey will result in the first comprehensive manual of aquifer properties and will be extended into Scotland through the involvement of the Scottish Office.

The improved operational management of rivers for flood defence purposes has been another key achievement of the R&D Programme. Work in this area has fed into an operational manual which is now in use by NRA staff, and covers river maintenance techniques, asset management and methods for prioritising programmes of channel maintenance.

The development of practical tools such as electrofishing boats, hydroacoustic survey methods and devices for tracking fish, have, and continue, to feature heavily in the R&D Programme. The use of electrofishing techniques, feature on BBC's Tomorrow's World, provides NRA staff with a safe method of stock assessment in some of the larger rivers under NRA control.

The development of new survey methodologies for assessing the conservation status of our rivers has enabled NRA staff to draw together clearer pictures of areas where habitats require protection or rehabilitation.

The Government has proposed an Environment Agency which will take over the responsibilities of the NRA, Her Majesty's Inspectorate of Pollution and the waste disposal duties of local authorities. The development of an integrated R&D Programme to support the duties and functions of the new Agency represents a significant challenge, and one in which the NRA will play an important role.

The NRA has continued to work closely with both the European Commission and organisations in individual Member States. Advice has been provided to staff of the European Commission concerning priorities for the Fourth Research and Technological Development Framework Programme, in areas such as flood warning and waste minimisation. Collaboration with the Po River Authority in Italy continues to provide a useful exchange of practical experiences on monitoring water quality.

This Annual R&D Review - 1994 also summarises the key outputs from the NRA's 1993/94 R&D Programme as well as providing details of projects in the 1994/95 Programme.

# 1. INTRODUCTION

The National Rivers Authority (NRA) is pleased to publish this, its fifth annual review of its research and development (R&D) programme.

## 1.1 This Review

### 1.1.1 Its Purpose

The NRA has a statutory duty under the Water Resources Act 1991,

*"to make arrangements for the carrying out of research and related activities (whether by itself or others) in respect of matters to which the functions of the Authority relate."*

In publishing this document, therefore, the NRA aims to provide a formal record of its progress in discharging this duty.

The NRA also has a stated aim of operating in an open manner, and is covered by the EC Directive on Freedom of Access to Information on the Environment. Given this climate, the NRA will continue to publish as much information as is practicable about its R&D Programme. In the majority of cases, the findings of R&D need to be put in the context of the NRA's activities and policies. This will inevitably result in a delay between a project being completed and the associated output being published.

In essence, this Review provides broad level information on the projects within the R&D Programme as well as the outputs produced. In this respect, this fifth review covers the period from April 1993 to March 1994.

This Review also serves to keep those many organisations with whom the NRA collaborates up to date on the rationale behind its R&D Programme, progress on projects and availability of outputs. This is essential if research-commissioning organisations, active in the water environment, are to work together in developing effective partnerships.

### 1.1.2 What Does it Contain?

The success of any review spanning a wide area of interests is often determined by the ease to which readers can "dip-in" to any subject area. With this in mind, the details of outputs produced and projects within the current R&D Programme are set out for each of the NRA's functions.

The overall structure of the Review is as follows:

- Section 1: the purpose of this Review, the challenges and opportunities, the strategy behind the programme, and how the NRA addresses issues such as partnerships, technology transfer and international links;

- Section 2: provides a self-contained review of the key aspects of the NRA's R&D Programme in its first five years of operation;

- Section 3: covers the rationale, programme and outputs for water quality-related R&D;

- Section 4: covers the rationale, programme and outputs for water resources-related R&D;

- Section 5: covers the rationale, programme and outputs for flood defence-related R&D;

- Section 6: covers the rationale, programme and outputs for fisheries-related R&D;

- Section 7: covers the rationale, programme and outputs for recreation - and navigation-related R&D;

- Section 8: covers the rationale, programme and outputs for conservation-related R&D;

- Section 9: covers the rationale, programme and outputs for cross-functional R&D;

- Appendix 1: provides a full listing of the projects undertaken in the NRA's 1994/95 R&D Programme;

- Appendix 2: provides other management information including personnel involved in R&D; and

- Inside of back cover: provides a full list of R&D outputs produced through the NRA's R&D Programme.

## 1.2 Challenges and Opportunities

### 1.2.1 The Environment Agency

The Government's proposals to establish an Environment Agency for England and Wales offer an exciting opportunity for focusing environmental research on providing useful management information and methodologies in order to improve the environment. The R&D Programmes of the NRA, Her Majesty's Inspectorate of Pollution (HMIP) and some pollution and waste-related areas of the Department of the Environment (DoE) will be brought together under an integrated framework capable of addressing significant issues such as catchment planning, best practical environmental option and sustainable development.

The long-term planning required for a successful R&D Programme has been a key factor behind the harmonisation of these programmes under the aegis of DoE's Environment Agency Advisory Group. The activities associated with this harmonisation are addressing the challenge of developing a business-driven, output-oriented R&D Programme, capable of drawing upon the best available scientific and technical expertise. This will be essential if the proposed Agency is to tackle the difficult problems ahead.

### 1.2.2 The EU Fourth RTD Framework Programme

The NRA, as the foremost environmental regulator in Europe, has taken a proactive stance in the formulation of priorities for the water-related elements of the EU's Fourth Framework RTD (Research and Technological Development) Programme. In October 1993, the NRA co-sponsored, with TechWaRe and the CEC, a Euroworkshop designed to discuss issues relating to water policy and to identify related future R&D requirements. The findings of this event were pulled together with issues identified by other Euroworkshops in Brussels, Lisbon and Potenza and were fed through to CEC staff. Many of the priorities are now reflected in the EU's specific programme on the environment and climate.

In response to the publication of the draft specific programme on the environment, the NRA has identified a series of priority areas where it will back focussed submissions. These areas have been discussed with some of the foremost research bodies in the UK with a view to establishing a coordinated position.

In taking forward proposals for EU funding, the NRA hopes to influence not only the research, but also the development of environmental policy which will undoubtedly benefit from such research in the future.

## 1.3 Strategy for Science and Technology

The NRA has welcomed the Government's White Paper "Realising our Potential" and the important emphasis it places on the creation of wealth and the quality of life. In particular, the NRA is keen to build upon the principles, set out in the White Paper, of providing an end-user focus for the scientific research carried out by Research Councils and academia in general.

The NRA R&D Strategy, published in 1993, sets out for each of its functions, the scientific rationale behind its R&D requirements. As a general principle, the NRA seeks to ensure that its decisions on environmental management are based on the best scientific information available. This is often achieved through reviewing recent advances in certain areas, or through the scoping of particular issues and assessing the most efficient and effective programme of work required to produce the outputs required.

A prime example of the NRA's approach to science and technology is the development of a strategic programme for groundwater research. In collaboration with British Geological Survey (BGS) and Foundation for Water Research (FWR), the NRA is seeking to scope out the issues associated with the management of groundwaters in England and Wales. Once this has been completed, the current state of knowledge in this area will be assessed and a collaborative programme of work developed to provide the management outputs needed.

In focussing its R&D requirements towards the applied and developmental end of the research spectrum, the NRA builds upon basic R&D funded research elsewhere. However, the NRA recognises that it is important to stimulate the science base in areas where more practical outputs may be needed in the medium term. With this in mind, the NRA has been funding a series of research fellowships.

Table 1.1 sets out the current position with regard to these important areas of research.

## 1.4 Partnerships in Science and Technology

The effectiveness of the NRA's R&D Programme has been enhanced by the level of collaboration and partnerships established with other research-commissioning bodies. In the five years of operating its programme, the NRA has worked closely with over 35 organisations, many of whom have brought not simply financial resources but also important technical expertise as well as a variety of viewpoints.

The issues involved in the management of the water environment are sufficiently complex for the development of partnerships to be essential if they are to be addressed

## Table 1.1 NRA Fellowships

**Methodologies for optimising the value of river corridor survey data**
Dr Peter Edwards, Geodata Institute, Southampton University

This Fellowship concentrated on the durability and utility of methodologies used for river corridor surveys. The six principal conclusions were:

1. River corridor survey maps provide a wealth of information, and have improved significantly.
2. In hard copy form, river corridor survey information does not permit overviews of rivers to be easily obtained.
3. Errors in locating features can be reduced to 2% through recording on undistorted maps and the use of aerial survey methods.
4. A balance needs to be struck between the clarity and the utility of the maps produced.
5. A six point abundance scale can be used to extract semi-quantitative information from the data.
6. Potential exists to integrate river corridor survey information with other data sources.

The findings of the Fellowship are set out in R&D Note 271. A number of the above recommendations are currently being taken forward by the Conservation function.

**Modelling the relative impact of weather, land-use and groundwater abstraction on low flows**
Dr Robert Wilby, Loughborough University

As part of the research undertaken through this Fellowship, computer software was developed for discerning the relative significance of the key factors associated with weather patterns.

In order to assess the potential impact of weather on flow regimes, a realistic climate scenario was constructed. Historical relationships between the prevailing Lamb Weather Type from 1861 to the present day and the likelihood of precipitation, were produced. This approach preserves the physical realism of the simulated rainfall regime at an appropriate temporal resolution for detailed impact assessments.

Following the development of this scenario, the hydrological response of the catchment was investigated using a semi-distributed, conceptual model which calculates total surface run-off.

The third stage involved the selection and analysis of model outputs to reveal the quantitative significance of each boundary condition. The research has several applications including statements of hydrological sensitivity to climate change and naturalisation of existing flow series.

This Fellowship has completed, and the results are summarised in R&D Note 268.

**Survey methodology for algae and other phototrophs**
Dr Martin Kelly, Durham University

The methods for sampling phototrophic organisms have been studied in a range of flowing water sites in the UK in order to assess their value as monitors of environmental change. The work has concentrated on epilitic diatoms as indicators of pollution.

Four diatom-based pollution indices have been examined. Two pollution indices (SPI and GDI) provided higher scores at upland sites even in the absence of organic pollution, whilst trophic diatom indices (TDI) gave higher scores at lowland sites. The TDI was correlated with filterable reactive phosphorus (FRP) but some modification will be necessary before it can be used in the UK. The GDI was also highly correlated with FRP and may also provide a good starting point for the development of indices for monitoring eutrophication.

A case study in the River Avon in Warwickshire was carried out to test the application of plant-based methods for monitoring related to the Urban Wastewater Treatment Directive. Estimates of the abundance of *Cladophora glomerata* were also recorded.

This work is continuing with a view to providing a trophic diatom index for assessing the trophic status of rivers. This may then be used as a monitoring tool for determining the impact of sewage discharges on the receiving waters. Progress to date is reported in R&D Note 278.

**The sustainable management of the water cycle**
Dr Richard Dubourg, Centre for Social and Economic Research on the Global Environment

This Fellowship has investigated the implications of sustainable resource management concepts for the management of water resources in England and Wales.

Freshwater may be regarded as "critical capital", and as such would lead to the requirement to manage this capital in a sustainable manner. In operational terms of course, the relevant water legislation, such as the Water Act 1989, sets out the task of sustainable water management.

In defining water as a capital item the physical characteristics of water may be regarded as policy control variables. The general sustainable development principles adopted for this Fellowship were that water abstraction should be met by net run-off only, and that water quality must not deteriorate over time.

Sustainable water use can be put into practice by ensuring that licensed net abstraction does not exceed gross regional resources and that the impact of discharge consents on water quality should not increase

The economic results of this Fellowship have been published through the Centre.

### Genetic structure of sea trout populations in England and Wales
### Dr Heather Hall, University of London

The work involved in this Fellowship is still on-going, and will complete in 1995.

A total of 1244 samples has been taken from sea trout and brown trout from a total of 123 river systems in England and Wales.

The analysis of mtDNA from the samples has revealed a clear divergence of populations with most falling into one of two groups. The genetic distance between populations within each group is low which points to a relatively recent divergence in sea trout populations. This divergence probably took place around 12 000 years ago.

The genetic data from this work has indicated that stocking has had a significant impact on the genetic diversity within some river systems. Sea trout from the rivers Coquet and Fowey have been identified to be particularly important for conservation.

---

effectively. The White Paper on Sustainable Development highlighted the success of the Aire and Calder project in bringing together a range of organisations to tackle waste minimisation and water conservation. In addition to the financing of the project, the organisations have established a good working relationship which has enabled other ideas and working practices to be discussed.

The Government's "Forward Look of Government-funded Science, Engineering and Technology" reinforced the need for more partnerships, particularly between the public and private sector, in R&D initiatives. The NRA aims to play a full role in this sphere where it is appropriate to do so.

In order to formalise some of these links, the NRA has established Memoranda of Agreement with both English Nature and the Scotland and Northern Ireland Forum for Environmental Research (SNIFFER). Discussions have also been held with the Natural Environment Research Council (NERC) with a view to establishing a concordat in line with the White Paper "Realising our Potential".

## 1.5 International Dimension

The remit of the NRA extends solely to England and Wales. However, it is also the competent authority for over 20 EC Directives and is often asked to provide advice in both European and world fora. In this respect, R&D is no different as new science and technical developments arise from various parts of the world. As summarised in Section 1.2.2 above, the NRA has been particularly active in the EU's Fourth Framework RTD Programme.

The NRA also cofunds a number of EU projects, and through such work NRA staff have an opportunity to influence the direction not only of the research, but also the eventual use to which the outputs will be put. An example of this has been the involvement of NRA staff at the SETAC World Conference in Lisbon in 1993, where the application of sediment ecotoxicological procedures and their link to NRA operational issues were discussed.

The NRA has also established working links with individual organisations across Europe, ranging from the Po River Authority in Italy to CEMAGREF in France. The aim of these links is to ensure that the NRA can benefit from the best practice available and not just experiences from England and Wales.

## 1.6 Dissemination and Technology Transfer

The benefits of R&D can only be realised if the outputs are transferred to those staff who require them and in a form that enables them to be easily used. The NRA has made major advances in the dissemination of R&D findings both to its staff, and to the general public.

Copies of all outputs made available to the general public are placed in the British Lending Library. Access to these outputs is now possible through any public or academic library. They are also available for inspection at the NRA's Information Centre in Bristol.

All outputs produced through the NRA's R&D Programme are now accompanied by detailed implementation plans, which set out the additional resources, or changes in working practices required, to put them into operation.

## 1.7 Operational Investigations

The NRA's R&D Programme is undertaken to provide information, tools and new techniques for all NRA Regions. However, there are projects which are targeted at Regional or site-specific problems. These are termed Operational Investigations (OIs), and tend to be less strategic, and short-term in nature.

There are a number of OIs that have been brought into the national R&D Programme in order to ensure that all Regions can benefit from the generic information developed.

A list of outputs produced through OIs is available from NRA Head Office.

## 1.8 The NRA's R&D Programme
### 1.8.1 Management Structure

The structure of the NRA's programme in commissions reflects its seven core function areas, plus the need to undertake some R&D on a multi-functional basis. Each R&D commission is strongly linked to the business programme of the core function concerned in terms of development of new policies, operational tools and knowledge to support the core function in carrying out its work in a more efficient and effective way.

The overall programme of work in each commission is supervised on behalf of the research customers by the so-called commissioner, who is a senior officer within the core function concerned. The commission programme is subdivided into a number of topic areas, each addressing a distinct theme within the business programme of the core function. Each topic area programme is developed and supervised by a topic leader having particular knowledge and expertise in the area concerned. Individual projects are supervised by project leaders. Both topic and project leaders are generally core function staff having operational or policy responsibilities in the areas addressed by their research interests.

The success of the NRA's R&D Programme is particularly dependent on both the ability and the contribution of these staff in achieving a satisfactory interface between operational and policy issues on the one hand, and the science base of research on the other. A list of staff involved in the 1993/94 programme is given in Appendix 2.

This management structure is more fully described in "The NRA's Research and Development Programme - A Position Paper by the Chief Scientist" which is available from the NRA.

### 1.8.2 Research Contractors

The NRA is also heavily dependent on the quality, skill and knowledge of its research contractors in delivering R&D outputs which are able both to address its business needs and to build on an adequate understanding of the science base.

In some areas, NRA research contractors are unique in their understanding of the issues concerned and in their nationally-acknowledged position. In most other projects, the NRA's research contractors are selected through the process of competitive tendering for specific research projects. This overall process of registration, preselection and tendering is described in the Chief Scientist's position paper referred to in Section 1.8.1.

### 1.8.3 Funding

The funding of the NRA's R&D Programme is dictated to some extent by the issues being addressed by the range of projects underway. In general, water quality, fisheries, recreation, navigation and conservation R&D projects, are resourced from grant-in-aid. The remainder, water resources and flood defence, are funded from their own ring-fenced accounts.

Cross-functional projects are cross-charged to the function in direct proportion to the relative benefit to each function.

Figure 1.1 illustrates the breakdown of the 1994/95 R&D Programme by type, or Frascati category. In addition, the programme is also categorised by the primary purpose of each project; these are summarised in Figure 1.2 for 1994/95.

Whilst the actual and planned expenditures for each Commission programme are provided in following sections, Figure 1.3 provides an overall summary of the actual and planned expenditures from 1992/93 until 1995/96.

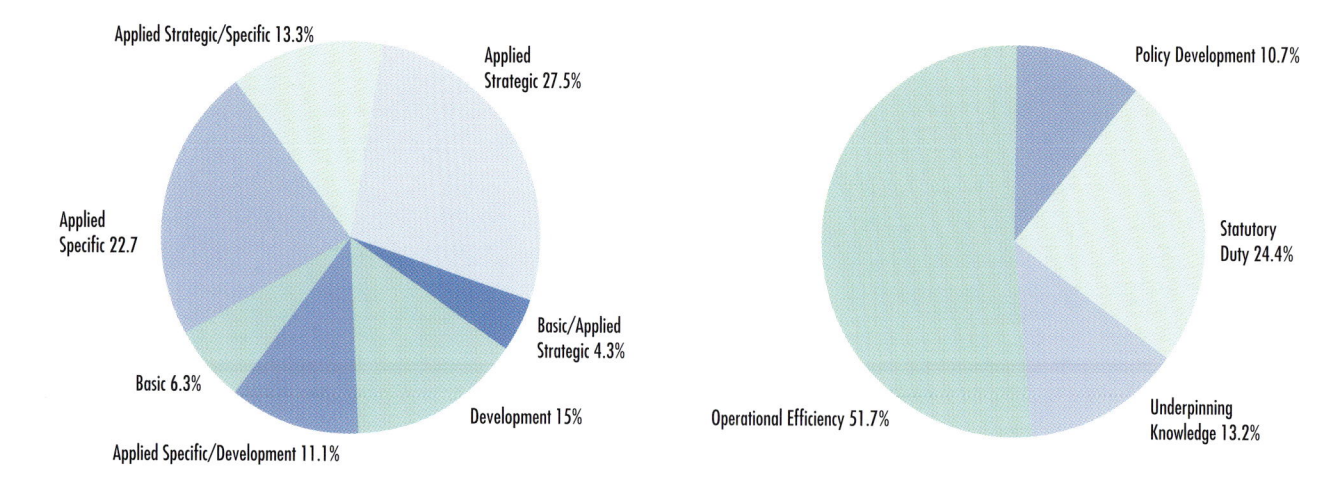

**Figure 1.1 Type of R&D in the 1994/95 Programme**

- Applied Strategic 27.5%
- Applied Strategic/Specific 13.3%
- Applied Specific 22.7
- Basic 6.3%
- Applied Specific/Development 11.1%
- Development 15%
- Basic/Applied Strategic 4.3%

**Figure 1.2 Primary purpose of R&D in the 1994/95 programme**

- Policy Development 10.7%
- Statutory Duty 24.4%
- Underpinning Knowledge 13.2%
- Operational Efficiency 51.7%

Legend: Water Quality, Water Resources, Flood Defence, Fisheries, Recreation and Navigation, Conservation, Cross-functional Issues, Management

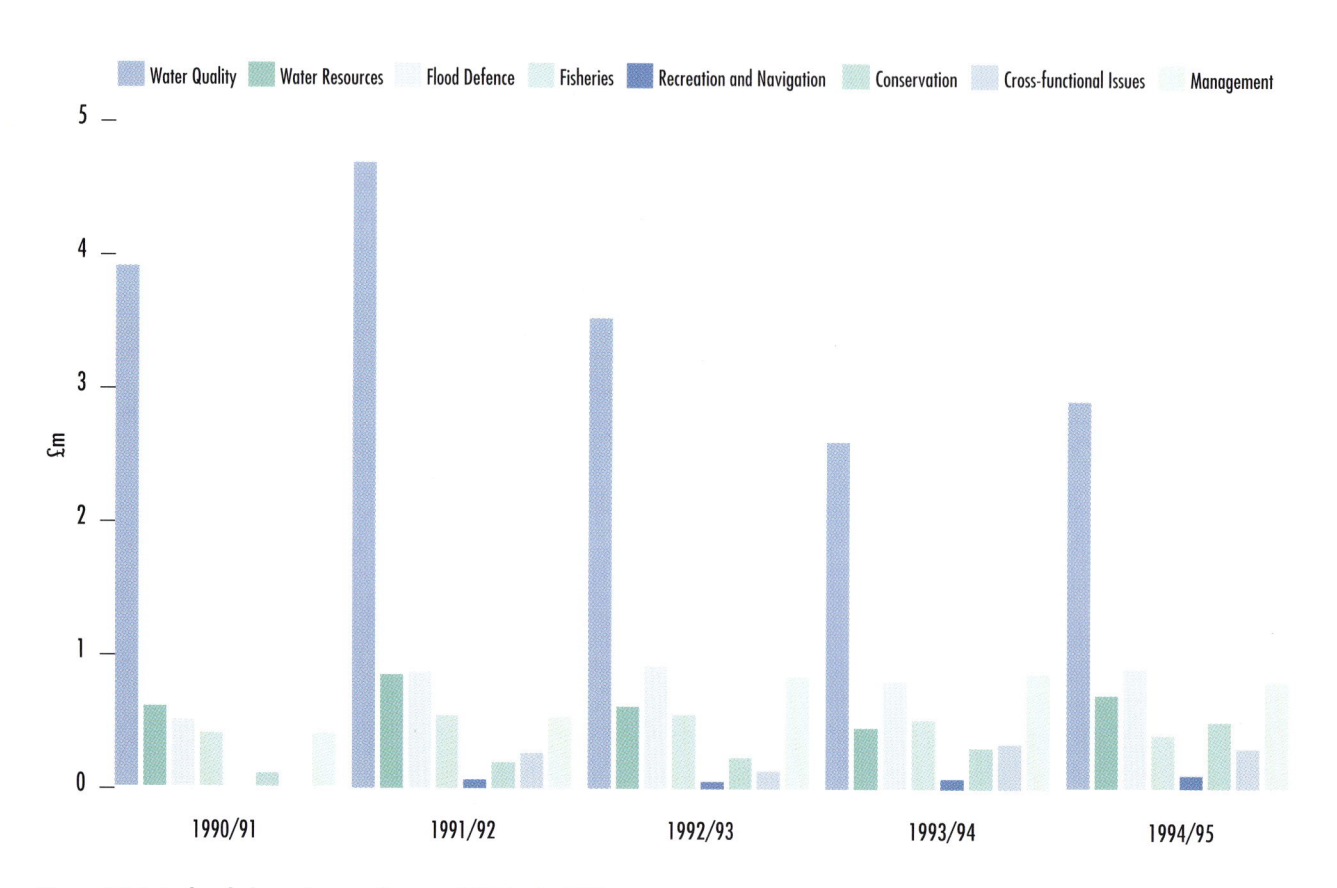

**Figure 1.3 Actual and planned expenditure on R&D in the NRA**

Annual R&D Review - 1994

## 2. REVIEW OF R&D IN THE NRA

This section provides a summary of the major R&D achievements and outputs of the first five years of the NRA. It highlights examples of significant projects carried out in support of the NRA's functions which have had an impact on the efficiency and effectiveness of day-to-day operational activities.

### 2.1 Introduction

Over the first five years of the NRA, functional priorities together with the public's overall perception of environmental issues have changed. As a support service, R&D has adapted to meet these changing requirements.

In moving from an inherited R&D Programme, the majority of which was specified before the NRA was envisaged, the NRA has ensured that R&D has become more business-driven. Key facets of the change include:

- function staff specify and subsequently manage the R&D;
- outputs are clearly defined at the outset of any project; and
- all outputs are accompanied by an implementation plan.

The process by which R&D is developed and managed is set out in Figure 2.1.

In introducing such changes, the NRA has ensured that the take up of R&D results has significantly improved, thereby improving value for money.

A further enhancement of the R&D Programme has been the increase in the number of beneficial collaborative initiatives. The NRA has been keen to develop partnerships wherever possible and has collaborated with over 35 different research-commissioning organisations, increasing the value of its programme by nearly 30% in some years.

The introduction of practical outputs from R&D projects has been accompanied by both tangible and intangible benefits. For instance, the NRA now holds more information on the quality of its coastal waters than any other EU member state through introducing remote sensing techniques. In addition, the introduction of instrumentation for monitoring water quality has helped reduce the expenditure on sample analysis.

In operating in a constantly changing physical environment, the NRA has benefited from a statutory duty to carry out R&D and to implement the outputs that have arisen as a result of its R&D Programme. A similar provision will be required by the proposed Environment Agency if it is to react to increasing pressures across an even wider spectrum of issues.

Set out in subsequent sections are examples of R&D projects where the information or equipment produced has improved the efficiency or effectiveness of the NRA.

Figure 2.2 illustrates the number of new projects developed over the first five years of the NRA's R&D Programme.

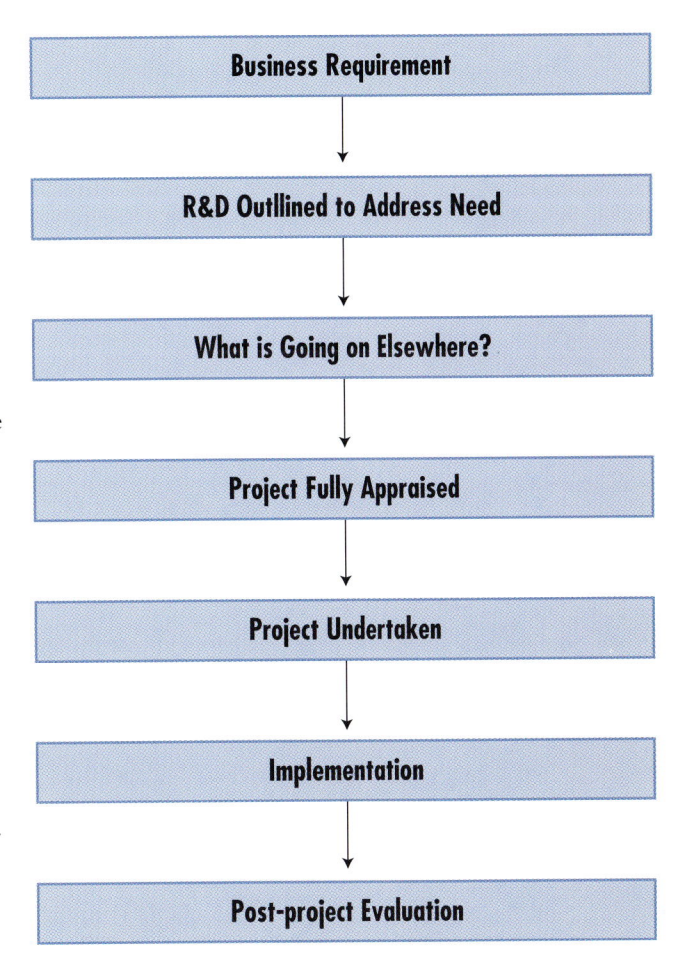

```
Business Requirement
        ↓
R&D Outlined to Address Need
        ↓
What is Going on Elsewhere?
        ↓
Project Fully Appraised
        ↓
Project Undertaken
        ↓
Implementation
        ↓
Post-project Evaluation
```

**Figure 2.1 The process behind the R&D Programme**

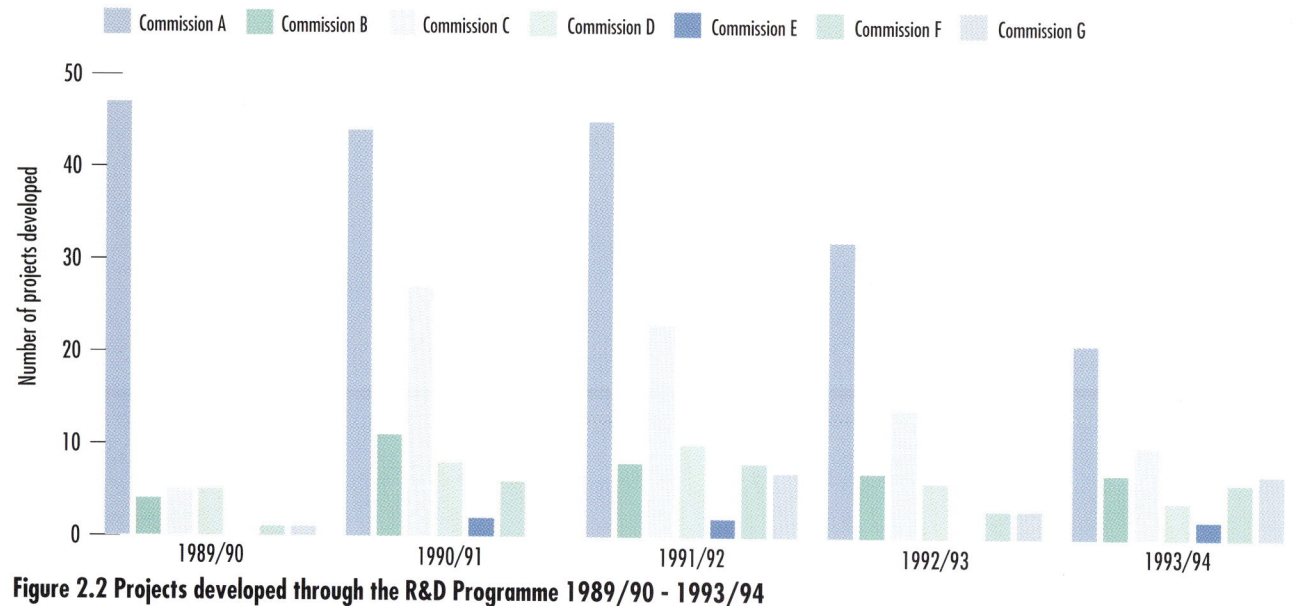

**Figure 2.2 Projects developed through the R&D Programme 1989/90 - 1993/94**

## 2.2 Waste Minimisation on the Aire and Calder

The industrial revolution and subsequent urban growth have left many great rivers with a legacy of serious pollution. However, the UK's first major study of waste minimisation as a means of reducing aquatic pollution and resource use, and cleaner technology looks set to become the catalyst for better "housekeeping" measures to reduce effluents and water usage, to collect significant financial savings, and to improve the quality of some important rivers.

The rivers Aire and Calder were typical of the industrial rehabilitation problem facing the NRA. However in collaboration with HMIP, BOC Foundation for the Environment and Yorkshire Water Services, a project was established to minimise the wastewater and the use of water in industrial processes.

The clear message of the project is that 'Pollution Prevention Pays' for both the environment and industry. During the project's first 18 months, the eleven participating companies made savings of over £2M a year; dispelling the myth that the reduction in pollution and improvements in profitability are mutually exclusive.

In "This Common Inheritance - Third Year Report", the Government has highlighted this project as an example of how partnerships between public and private sector can play a significant role in the sustainable development of the environment.

*NRA Project Leader:*    *Dr Tony Edwards, Northumbria & Yorkshire Region*
*Research Contractor:*    *Centre for Exploitation of Science and Technology*

## 2.3 Cyclops - Automatic Formal Sampling Machine

Whilst continuous water quality monitors have been in use in the NRA for a while, their use in the consent compliance process has been limited due to the need for tripartite samples for prosecution. Cyclops has been developed to fulfil this need, combining the most modern sensors with responsive sampling equipment.

**Cyclops - continually protecting the river**

The monitor assesses levels of pollutants in the river adjacent to a discharge. When preset levels are exceeded, a telefax is sent by Cyclops over the cellular telephone network to the local pollution control office. If required, the NRA pollution staff can prompt the machine to take a formal sample which, when opened in the presence of a representative of the discharger, can be used for prosecution purposes.

This R&D output has already been demonstrated in practice, providing information for the successful prosecution of a major UK chemical company for breaching the pH consent value.

Cyclops is one of a range of water quality instruments being employed by the NRA's National Centre on Instrumentation and Coastal Surveillance in the fight against pollution.

*NRA Project Leader:* Paul Williams, South Western Region

*Research Contractor:* In-house

## 2.4 Flood Defence Strategic Planning

Control of development in flood risk areas is part of the NRA's Flood Defence Strategy. Effective control should ensure that development does not worsen the risk of flooding due to either increased run-off from urban development, which can give rise to flooding elsewhere, or to construction in the floodplain, which is in risk of flooding and/or reduces flood discharge capacity. The NRA is a consultee in the planning process and plays a key role in planning to prevent flood risks.

**Aerial photograph of urban flooding**

Three projects - (a) Strategic Approach to Planning and Flood Risk, (b) Best Practice in European Strategic Land Use Planning and (c) Efficiency and Effectiveness of Planning Activities - have identified key features of best practice which will enable the NRA to adopt a more proactive and planning approach to development control.

The project recommendations have contributed to new guidelines for NRA input to local authority development plans. The NRA will also be producing a national Policy for the Protection of Floodplains, and achieving closer and more effective communication with planning authorities through its Memorandum of Understanding and Guidance Notes for Local Planning Authorities.

Many of these measures are supported by the emphasis given in the Government's recent White Paper on Sustainable Development to increase community involvement, partnerships, and public consultation in environmental management.

*NRA Project Leaders:* Tony Burch, South Western Region (a)
Barry Winter, Thames Region (b)
Alan Hopkins, Southern Region (c)

*Research Contractors:* Middlesex University (a) and (b)
M Parker Associates (c)

## 2.5 Electrofishing and Hydroacoustic Surveying

An important activity carried out in support of the NRA's duty to maintain, improve and develop fisheries, is stock assessment. Two R&D projects have provided equipment to enable NRA staff to collect information on fish stocks across a wide range of rivers whilst at the same time avoiding causing harm to the fish themselves.

The first of these projects has produced an electrofishing boat capable of being used safely in large yet shallow lowland rivers.

**Electrofishing as part of routine stock assessment**

The boat applies an electric current to the water, stunning the fish which rise to the surface and can then be analysed before being returned safely to the river. The use of this equipment was highlighted on BBC's Tomorrow's World.

The second project has further developed the traditional echo-sounder in order to provide a three-dimensional picture of fish distribution in a river. The information returned to either the boat or the river bank can be used to estimate fish biomass and, in some cases, fish species.

These two projects have produced operationally useful items of equipment which enable NRA fisheries staff to assess fish stocks more effectively.

*NRA Project Leaders:*    Dr Phil Hickley, Severn-Trent Region (Electrofishing)
     Dr Alan Butterworth, Thames Region (Hydroacoustic)
*Research Contractors:*    Humberside International Fisheries Institute (Electrofishing)
     Royal Holloway and Bedford New College (Hydroacoustic)

## 2.6 Low Flows and PHABSIM

The adverse impact of low flow conditions on the environment of many UK rivers has provided the impetus for an important R&D project.

The NRA has a stated aim of determining Minimum Acceptable Flows and alleviating low flows where possible. In doing so it has identified a need for further information on the flow requirements of various aquatic flora, fish, invertebrates and macrophytes.

The model that has been developed, Physical Habitat Simulation System (PHABSIM) has been used in the establishment of minimum acceptable flows in rivers, as well as for identifying where existing low flows may be causing problems for the ecology.

In order to validate the model, which is now in use within the NRA, habitat preference curves were produced which involved NRA staff snorkelling alongside trout and salmon to see how they reacted to changes in flow.

The pioneering application of these methods in the UK to the River Allen problem gives grounds for real optimism that at last there is a technique through which to define quantitative measures of ecology on a scientific basis.

*NRA Project Leader:*    Terry Newman, South Western Region

**Alleviating low flows as part of protecting the environment**

*Research Contractors:*    Institute of Hydrology and Institute of Freshwater Ecology

## 2.7 The Properties of Groundwater Aquifers

As part of the NRA's broad strategy for Water Resources, it aims to plan for the sustainable development of water resources, developing criteria to assess reasonable needs of abstractors and of the environment.

In planning for such development of groundwaters, the availability of comprehensive and scientifically sound information on the properties of aquifers is of paramount importance.

The NRA has been working in collaboration with the BGS to produce a definitive guide on the properties of all aquifers in England and Wales. This has been extended to Scotland through collaboration with Scottish Office Environment Department.

The manual itself will be an essential tool in evaluating the yield of aquifers and the need for groundwater protection, as it will provide information on the potential sensitivity of aquifers

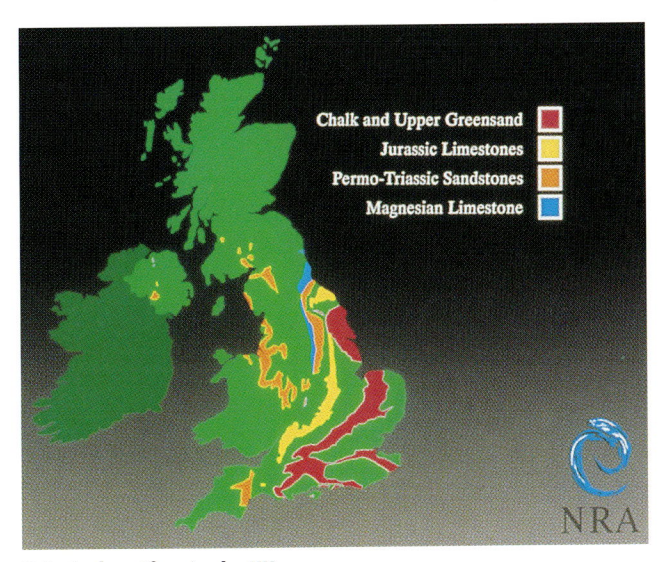

**Principal aquifers in the UK**

above which developments are planned. This manual will be published in 1996.

*NRA Project Leader:* Vin Robinson, Thames Region
*Research Contractor:* British Geological Survey

## 2.8 Remote Sensing Techniques

The NRA has always been keen to keep abreast of new innovative tools and high-tech methods which help in its day-to-day activities. Remote sensing was one of these techniques targeted early on and a cross-functional review of available sensors both satellite-based and those flown in aircraft was undertaken.

The review identified the benefit of visible sensors such as the Daedalus ATM and the Compact Airborne Spectral Imager (CASI). A full-scale field comparison of these two sensors was carried out to assess whether the logistics of operating such instruments together with supporting survey vessels was cost-effective for water quality surveillance.

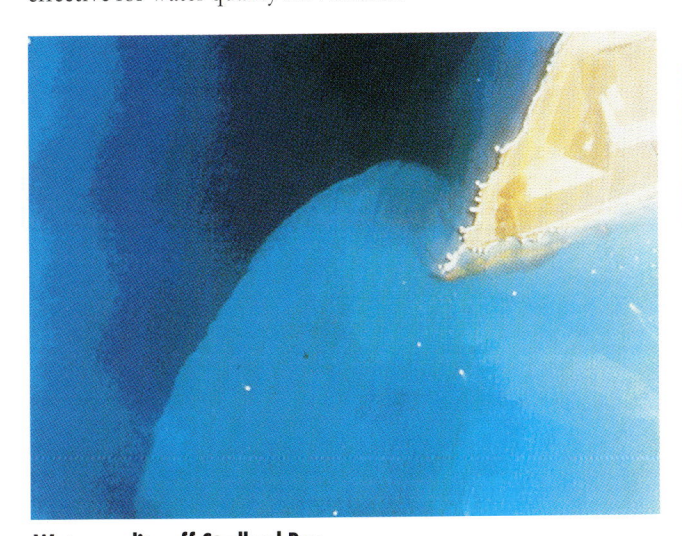

**Water quality off Studland Bay**

The CASI is now in operational use within the NRA as part of the National Centre on Instrumentation and Coastal Surveillance. It is used to provide synoptic information on the quality of 20,000 km² of coastal waters of England and Wales.

This technique has given the NRA more information on the coastal waters of England and Wales than ever before. It has enabled better targeting of more detailed local surveys.

The Government's White Paper on "Sustainable Development" has highlighted the success of the NRA's activities in this field.

Further collaborative R&D is underway with NERC to use the information collected by the CASI for river corridor conservation surveys, monitoring beach movements for Flood Defence and for assessing the quality of rivers and lakes.

*NRA Project Leaders:* Gareth Llewellyn,
Head Office (Review)
David Palmer,
South Western Region
(Comparison)
Nick Holden,
South Western Region
(Further R&D)
*Research Contractor:* Natural Environment
Research Council

## 2.9 Flood Defence Operational Management

Substantial resources are needed for Flood Defence which accounts for the highest expenditure of all the NRA functions. It is essential that these activities are carried out economically, efficiently and effectively.

The overall objective of the R&D Programme in support of Flood Defence operational management has been to develop nationally consistent planning and management systems which ensures that Flood Defence operations, maintenance and capital

**Weed cutting to increase channel capacity**

works programmes throughout the NRA are consistent, prioritised, adequately justified and cost effective.

A report on Improving Efficiency and Effectiveness in Flood Defence Operational Management has brought together the key findings of the programme in:

- Standards of Service (SoS) for Flood Defence;

- Prioritisation and Programming of Flood Defence work;

- Asset Survey;

- Best Maintenance Practices; and

- System Performance Appraisal.

As R&D cannot exist in isolation from actual practice, the results of these projects have been incorporated into planning and management systems which have been developed by NRA staff, including the Flood Defence Management Manual which provides a systematic guide to data collection and analysis. The rewards of the R&D are now being realised as techniques are adopted at the grassroots level becoming common practice within the NRA.

*NRA Topic Leader:*    *John Fitzsimons, Severn-Trent Region*
*Research Contractor:*    *Mott MacDonald and Gould Consultants*

## 2.10 Conservation Handbooks

The effective management of the conservation resource and assessments of the factors affecting the habitats around rivers relies heavily on robust and consistent field information.

A number of projects have contributed to the development of handbooks for conservation staff to use in the field. These have included:

- river landscape assessment methods;

- river corridor survey methods;

- crayfish identification; and

- atlas of aquatic flora and fauna.

These methods are being developed to support the implementation of a conservation classification scheme, and ensure that NRA staff provide consistent levels of information for management decisions.

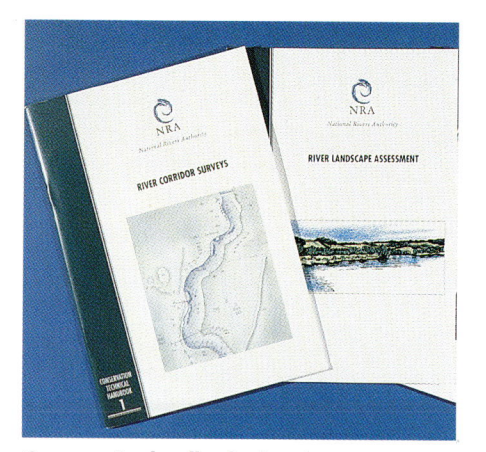

**Conservation handbooks drawing on up to date R&D**

In addition to these specific outputs, the NRA has also funded the development and publication of biological keys at the Freshwater Biological Association.

*NRA Project Leaders:*    *David Hickie, Severn-Trent Region (Landscape assessment) John Hogger, Northumbria & Yorkshire Region (Corridor survey, Crayfish and Atlas)*
*Research Contractors:*    *Land Use Consultants (Landscape Assessment) WRc (Corridor Survey) University of Nottingham (Crayfish) Institute of Terrestrial Ecology (Atlas)*

## 2.11 Environmental Economics

The NRA has a stated aim of demonstrating value for money whilst delivering its statutory duties. However, an assessment of the value that can be placed on the environment is particularly difficult. To advance this area of science, the NRA commissioned a number of studies to develop specific methods for allocating costs to particular environmental improvements. Some of these have been summarised in an NRA R&D Report entitled "The development of environmental economics for the NRA".

In addition to the NRA Fellowship on 'The sustainable management of the water cycle', the NRA is currently co-funding work with UKWIR, DoE and SNIFFER on the costs and benefits of river water quality improvements. The interim manual currently in use has set out a range of economic techniques for determining whether capital improvements at sewage treatment works can be balanced by the resultant environmental improvements.

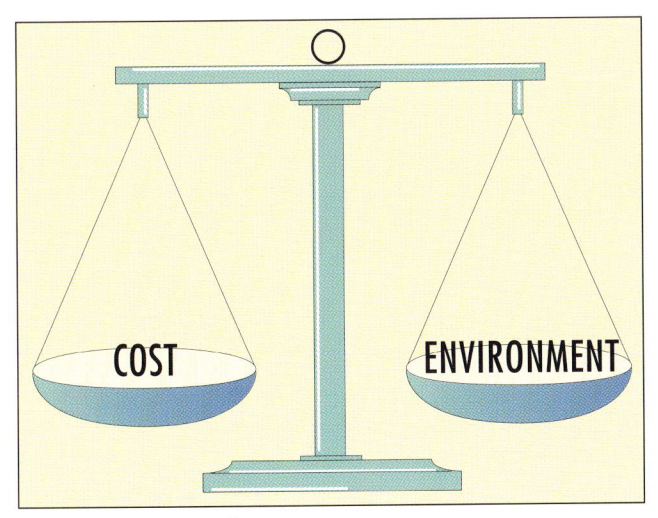

**Getting the balance right**

These techniques have also been developed to assess whether the benefits of alleviating low flows in certain rivers outweigh the costs of increasing the river flow.

Many of the techniques have fed into the development of an economic appraisal manual currently being used by NRA staff and further work will be carried out to enhance the methodologies.

| | |
|---|---|
| NRA Project Leaders: | Meg Postle, Economic appraisal |
| | Jerry Sherriff, Head Office (Low flows) |
| | Charlie Pattinson, Welsh Region (Water quality) |
| Research Contractors: | WS Atkins and Middlesex University (Low flows) |
| | WRc and Oxera (Water quality) |

## 2.12 Other Outputs

Over the first five years of operation, the NRA's Programme has produced in excess of 360 outputs. These are catalogued in the 'List of R&D Outputs' provided as an insert in the back of this document, and the numbers produced are summarised in Figure 2.3.

In addition to those covered in Sections 2.2 to 2.11, there are a range of other outputs which impact on the NRA's operations, namely:

- novel technology;
- working with others; and
- new use of existing technology.

Examples of the first of these categories include the use of miniaturised circuitry to provide personal alarms for NRA staff working alone in the field, and disposable sensors for monitoring metals in waters.

The partnership principle outlined in Section 2.2 is also clearly illustrated through the work on the urban pollution management manual and fish tracking developments where the NRA is able to work with other organisations active in a particular field.

The final category involves the exploitation of existing technology in a form which may be of use to the NRA. The use of fibre optics for the measurement of ammonia in freshwaters is a prime example of this approach.

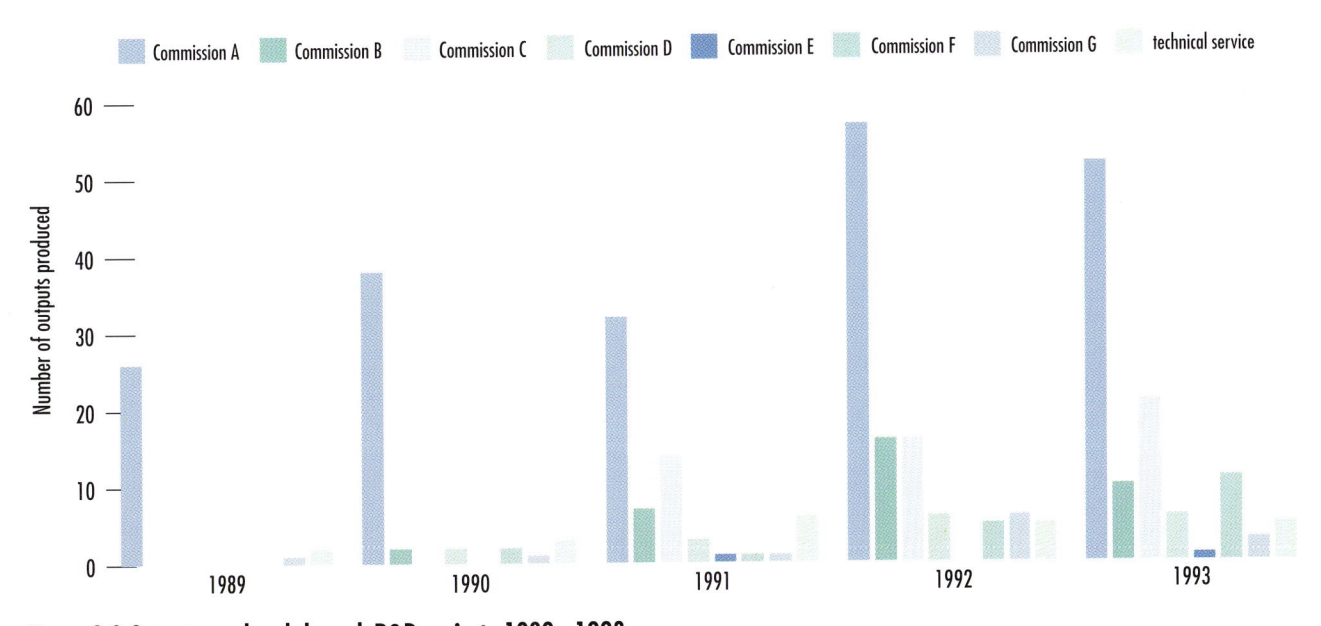

**Figure 2.3 Outputs produced through R&D projects 1989 - 1993**

Annual R&D Review - 1994

# 3. WATER QUALITY

This section sets out the progress made in the Commission A R&D Programme in 1993/94. Also described here are the Topic Programmes for the 1994/95 financial year, together with the strategic direction of the Commission A. This area of R&D is overseen by John Seager, Head of Environmental Quality at Head Office, as Commissioner.

## 3.1 Business Rationale

The NRA published its Water Quality Strategy in 1993, setting out its objectives and key activities over a ten year horizon. The key areas of work have been consolidated through the establishment of four business groups, namely:

- Statutory Water Quality Objectives (SWQOs);
- Water Quality Monitoring;
- Discharge Control and Charging; and
- Pollution Prevention.

All R&D carried out in Commission A provides support to these four business areas.

In addition to the development of SWQOs, the NRA is continuing the development of a General Quality Assessment Scheme (GQA) as the principal tool for assessing and reporting on the state of the water environment. This work will require significant R&D input, particularly for the new criteria such as nutrients and aesthetics.

The NRA has a stated aim of improving the efficiency of its activities, and this has been the driving force behind a review of the monitoring activities of the NRA. The introduction of new methods of working and innovation technology will assist in making water quality monitoring both more efficient and more effective.

The issuing of discharge consents is an important process in the NRA's drive to control point source pollution. The piloting of toxicity criteria within consent conditions could lead to an improved ability to control particularly complex effluents.

Pollution from diffuse sources, such as leachate from landfills and agricultural practices, is a continuing problem. The NRA has introduced a range of preventative measures to assist in the control of this type of pollution, but further R&D is needed to support activities in this area.

## 3.2 Outputs Produced

A key component of water quality management is the introduction of Environmental Quality Standards (EQSs). As part of the R&D Programme on EQS development, proposed EQSs were put forward for sheep dip chemicals such as chlorfenvinphos as well as for flue gas desulphurisation effluents.

A major report on the effects of sheep dip chemicals on water quality (R&D Report 11) was published in 1994. This work sets out the potential problems that may arise from the incorrect disposal of sheep dip waste and provides information for an informed debate on the subject.

The findings of a study assessing the levels of the dioxins PCDDs and PCDFs in surface waters were also released in 1994. The R&D provided information on levels across Europe and placed particular "hot spots" in their true context.

Work was also completed on the development of a quality system for water quality monitoring instrumentation. The output from this study will provide a useful basis for the work of the new NRA National Centre on Instrumentation and Coastal Surveillance.

## 3.3 Scientific Rationale

In common with the R&D Programmes of other Commissions, R&D undertaken for Water Quality is designed to provide the best scientific information and techniques in an easily useable form.

Outputs, such as the estuarine toxicity test using indigenous mysids, give NRA staff access to new scientific developments. The use of buffer strips in forestry activities as a method of protecting upland water quality has drawn upon new scientific research on the movement of pollutants through soils adjacent to small streams.

The disinfection of sewage effluents by many Water Service plcs has resulted in the NRA investigating the potential by-products of such techniques. An improved understanding of the chemistry involved as well as the microbiology of the organisms being targeted has resulted from new information in both areas.

## 3.4 Strategic Projects

Three very significant projects were started in 1993/94 which will provide a focus for efforts in three of the business areas listed in Section 3.1 above.

The development of a GQA scheme for rivers, estuaries and coastal waters was started in 1994 after a earlier scoping study. This new classification scheme will provide a central tool in the assessment of water quality for years to come.

An assessment of current best practice within the NRA followed by a similar review across Europe is the focus of the study on improved monitoring efficiency. This project will look at the introduction of new technology for planning more effective sampling strategies as well as the assessment of water quality itself.

The scoping of the R&D needed to pilot toxicity based consents was reported in the Annual R&D Review for 1993. The pilot study itself was started in 1993/94 and will be carried out in conjunction with both SNIFFER and HMIP. If successful, the results of this project could significantly enhance the NRA's ability to control point source discharges.

## 3.5 Topic Programmes and Outputs

This section covers the rationale behind the projects undertaken in each of the nine Topic Areas within the Water Quality Commission. The outputs that have been produced between April 1993 and May 1994 are also included.

Summary information is provided for those outputs available to the general public; a complete listing of all outputs together with their availability is provided as an insert in the back cover.

Where an output is still under consideration by the NRA, where it contains commercial in confidence information, or where it feeds into another activity such as the development of a functional manual, its current availability may be less widespread. External organisations wishing to know more about such outputs should contact the relevant Topic Leader.

The R&D Programme for each Topic Area is provided in Tables 3.1 to 3.9. These tables list the projects either on-going or proposed for starting in the financial year 1994/95. Each Regional R&D Coordinator and the R&D Section at Head Office have further details on all such projects.

A fold-out guide to the information contained in the following sections is provided on the inside front cover.

### 3.5.1 Standards and Classification Schemes

**Topic Leader:**
Mark Everard, Water Quality Planner with Head Office.

**Rationale:**
The research in this Topic is designed to support the development of both the SWQO scheme and the implementation of the General Quality Assessment Scheme. The work involves the provision of scientifically-sound environmental quality standards (EQSs) and a framework within which the water quality can be managed effectively. This Topic Area supports the activities of the SWQO project.

**Table 3.1 Standards and Classification Schemes (Topic A1) R&D Programme (1994/95)**

| Proposal/ Project No. | Title | Start/ End | Contractor | Comments |
|---|---|---|---|---|
| | **On-going projects A1** | | | |
| A04(91)7 390 | Joint Nutrient Study (JoNuS) To understand the scale, trends and processes of nutrient cycling in major east coast estuaries and coastal waters. | 11/91 1/96 | MAFF-led consortium | Collaborative project with MAFF |
| A04(92)2 457 | Ionisation of ammonia in estuaries To develop theoretical models which allow the concentration of toxic ammonia in estuarine waters to be calculated from a knowledge of salinity, temperature, pH and total ammonia. | 2/93 9/94 | PML | |
| A08(90)1 286 | Lakes - Monitoring and classification To develop an effective system of monitoring and classification of lakes and other standing bodies of water. Phase 2 - testing and evaluation of classification | 1/91 7/93 94/95 95/96 | University of Liverpool | Links with Projects 469 and 479 |

| Proposal/ Project No. | Title | Start/ End | Contractor | Comments |
|---|---|---|---|---|
| A09(90)1 227 | Metal speciation in rivers and estuaries<br>To determine the chemical form of selected metals in rivers and estuaries. | 7/92<br>5/93<br><br>6/94<br>3/95 | WRc | |
| 037 | Development of microbial standards<br>To define the possible effects of recreation by special interest groups in UK marine waters and to develop standards based on Phase 1. | 6/89<br>9/93 | WRc | |
| 053 | Development of Environmental Quality Standards, Phase 2<br>To develop environmental quality standards (EQSs) for substances of concern to the NRA. | 7/93<br>5/94 | WRc | |
| 051 | European pollution control philosophy<br>To examine European pollution control strategies to assist in the development of related UK initiatives. | 6/94<br>3/96 | WRc | Part-funded by SNIFFER |
| A(93)1 469 | General Quality Assessment Scheme<br>To develop and test a General Quality Assessment scheme for rivers, estuaries and coastal waters to provide an objective and comprehensive classification framework for river quality surveys. | 10/93<br>2/95 | WRc | |
| **Proposed new starts A1** | | | | |
| A01(94)1 | Toxicity-based criteria for assessing receiving water quality<br>To develop and assess toxicity-based criteria in order to assess the general quality of receiving waters - Phase 1 : scoping study for marine waters, to include a review of biomarkers. | 94/95<br>94/95 | | Links to Project 469 |
| A01(94)2 | National database on recreational freshwater quality<br>To assemble all available data on recreational freshwater quality to assist in the setting of SWQOs. | 94/95<br>94/95 | | Links to Topic E1 |
| A01(94)3 | Assessment of metals in marine sediments<br>To assess the level of metals and particle sizes within marine sediments to support the development of GQA scheme. | 94/95<br>94/95 | MAFF | Part-funded by MAFF |
| A08(94)5 | Faecal coliforms in shellfish and surrounding waters<br>To assess the link between bacterial levels in shellfish flesh and surrounding waters to control pollution. | 94/95<br>94/95 | | Part-funded by SNIFFER |

## Outputs

This section lists the principal outputs that have been produced through the Topic Programme in 1993/94.

| | |
|---|---|
| R&D Project Record 016/8/N | Investigation of partitioning between water and sediment |
| R&D Note 165 | Development of microbial standards |
| R&D Note 203 | Effects of sediment metals on estuarine benthic organisms |
| R&D Note 119 | Pathogenic microorganisms |
| R&D Note 188 | Correlation between enterovirus and faecal indicator organisms |
| R&D Project Record 411/3/Y | Correlation between enterovirus and faecal indicator organisms - Detailed statistical review |

# Summary Information

The following summaries are provided for those documents available to the general public.

**NRA Publication Ref: R&D Note 203**

**Effects of Sediment Metals on Estuarine Benthic Organisms**

Langston W J et al. (1994)

Plymouth Marine Laboratory

NRA Project No. 105

A major aspect of this research contract was to study the success of the legislation in some of the most heavily TBT-polluted areas of the UK including Poole Harbour and Southampton Water.

After 1987, seawater concentrations of TBT declined in small-boat areas but by 1992 were, at the more contaminated sites, still at least ten times the Environmental Quality Standard (EQS) of 0.8ng/l Sn (2ng/1 TBT). In other areas, TBT in sediments declined but at a much slower rate than in the water, for example in the Itchen Estuary, half-times for the decline of TBT in water and sediment were 3 years and 7 years respectively. Sediment particles scavenge TBT from the water column and provide a reservoir of the compound. Rather than reducing TBT bioavailability, sediments may provide a major route for uptake (and toxicity). This appears to be the case for the deposit-feeding clam, *Scrobularia plana,* which was common in the 1970s but had declined considerably by the 1980s. At less polluted sites there are now signs of recovery as TBT levels fall. The dog-whelk, *Nucella lapillus,* is so sensitive to TBT that surviving populations in the study areas still show no evidence of recovery.

Key words: Metal, pollution, tributyltin, mercury, methylmercury, copper, sediment, dog-whelk, mussel and bioavailability.

**NRA Publication Ref: R&D Note 119**

**Pathogenic Microorganisms**

1992

Centre for Research into Environment and Health

NRA Project No. 290

The study has defined algal, bacterial, protozoan and viral pathogens of concern to the NRA. It gives the best current information for each organism with regard to its occurrence, pathogenicity, sources, and longevity in the environment. It also gives brief details of the diseases attributable to them, and analytical considerations including a table of approximate costs of analysis.

In addition to providing information which enables the NRA to better discharge its statutory responsibilities for regulating water quality, the output from this project can also serve to direct future R&D projects in the microbiological field.

Key words: Pathogens, microorganisms, controlled waters, algal, bacterial, protozoan, viral, occurrence, sources, longevity.

**NRA Publication Ref: R&D Note 188**

**Correlation between Enterovirus and Faecal Indicator Organisms**

Wyer M D and Kay D (1993)

Centre for Research into Environment and Health

NRA Project No. 411

The report outlines the strength and nature of correlations between faecal indicator bacterial concentrations and enterovirus concentrations in seawater. A national database of microbiological water quality parameters (coliform organism, faecal coliform, faecal streptococci and enterovirus counts) from the Regional public register archives was developed. This database contains details of 114,362 bacterial and 3,015 enterovirus determinations.

The correlation analysis revealed low positive correlations between bacterial indicator enterovirus counts. Whilst the correlations were highly statistically significant, due to the large amount of data available, the capacity of indicator concentration to predict that of enterovirus was found to be uniformly low. There was no evidence of a level of bacterial concentration above which the correlation with enterovirus improved. Within the limits of the data available for analysis faecal indicator bacteria cannot be considered as useful predictors of enterovirus concentration in the coastal waters of England and Wales.

Key words: Viruses (enteric), bacteria (coliform), bacteria (faecal), coastal waters, correlation, regression analysis.

NRA Publication ref: R&D Project Record 411/3/Y provides a detailed statistical analysis in support of R&D Note 188.

### 3.5.2 Monitoring - Strategy and Reporting

**Topic Leader:**
Jacqui Gough, Water Quality Assessor with Head Office.

**Rationale:**
The main accent of the work in this Topic Area is to develop protocols and codes of practice for use by NRA staff involved in the design of surveys and sampling programmes. The R&D includes initiatives on the collection, preservation and storage of samples and the related data handling needs. This Topic Area will support the development of a national monitoring protocol as part of a functional strategic initiative.

**Table 3.2 Monitoring - Strategy and Reporting (Topic A2) R&D Programme (1994/95)**

| Proposal/ Project No. | Title | Start/ End | Contractor | Comments |
|---|---|---|---|---|
| | **On-going projects A2** | | | |
| A11(92)1 428 | Automatic exception reporting for trends in quality  To develop a semi-automatic exception reporting system, which detects trends in water quality data as soon as possible after their onset, to provide NRA staff with the means to make more effective use of all routine chemical monitoring data. | 11/92 10/93 | WRc | Part-funded by SNIFFER |
| 015 | Atmospheric inputs of pollutants into surface waters  To determine the organic composition of atmospheric deposition in the UK and assess the contribution of airborne pollutants to the organic content of surface waters. | 10/88 6/94 | WRc | |
| A07(92)9 449 | Modelling *E. Coli* in streams  To develop and apply a module for QUASAR to predict *E. Coli* in freshwater catchments. | 2/93 3/94 | IH | Contribution to DoE project |
| A04(92)5 458 | Estuarine water quality model  To develop, as part of the NERC LOIS study, a portable estuarine water quality model to enable classification of waters in relation to SWQOs. | 3/92 4/95 | NERC | Part-funded by NERC and DoE |
| A(93)2 479 | Improved monitoring efficiency  To undertake practical catchment-based case studies to review existing monitoring activities and to develop and demonstrate new monitoring programmes and associated data analysis tools which provide an integrated approach and value for money. | 4/93 7/93 1/94 3/95 | WRc | Collaborative project with Po River Authority, Italy |
| A11(92)2 482 | Early warning system for potential failure of river class  To develop a system that would routinely inform the NRA as soon as possible of any river sampling point that was in danger of failing its class. | 7/93 8/94 | WRc | Part funded by SNIFFER |
| | **Proposed new starts A2** | | | |
| A02(94)1 | Further testing and development of QUASAR  To further develop the river quality model QUASAR to enable NRA staff to use it for specific end-user applications. | 94/95 95/96 | IH | Contribution to NERC/LOIS project |

## Outputs

This section lists the principal outputs that have been produced through the Topic Programme in 1993/94.

| | |
|---|---|
| R&D Project Record 117/2/A | Environments of larger UK rivers |
| R&D Note 242 and R&D Project Record 330/4/A | Distribution of PCDDs and PCDFs in surface freshwater systems |
| R&D Note 183 and R&D Project Record 361/4/NW | Analysis, storage and archiving of water quality data |

R&D Note 121   Combined distribution method for estuary modelling

R&D Note 138   Quality control of sampling procedures

R&D Note 241   Codes of practice for data handling - Version 1

## Summary Information

The following summaries are provided for those documents available to the general public.

**NRA Publication Ref: R&D Project Record 117/2/A**

**Environments of Larger UK Rivers**

Reynold C S and Glaister M S (1992)

Institute of Freshwater Ecology

NRA Project No. 117

The report describes the outcome of a project to develop the understanding of the distribution and functioning of fluid dead-zones in rivers and to gauge the impact of retentivity on the structure and well-being of riparian communities.

The data set is introduced, outlining the choice of rivers and sampling sites, information on the plankton supported, preliminary data on down stream increase and an explanation of the derivation of growth rates and travel times.

Inferences about the velocity of flow at selected stations and through selected reaches, based on phytoplankton growth, are compared with traditional calculations. Some river-reaches are relatively retentive (i.e. they impede passage of water for longer than the apparent time of travel), mostly as a function of the reach gradient and sinuosity.

By reference to differences in phytoplankton abundance, both in space and time, at selected locations, the mechanisms of retention are explored.

The importance of patchiness in velocity structure to the maintenance of a potamoplankton is emphasised.

The use of phytoplankton as an index to river behaviour is not advocated as a technique but the study has shown that a response is demonstrable which may provide an explanation for the survival and (at times) prominence of river plankton.

Key words: Phytoplankton, dead zones, flow rivers.

**NRA Publication Ref: R&D Note 242**

**Distribution of PCDDs and PCDFs in Surface Freshwater Systems**

Rose C L, McKay W A and Ambidge P F (1994)

AEA Technology

NRA Project No. 330

The report determined the distribution of polychlorinated dibenzo-p-dioxins (PCDDs) and polychlorinated dibenzofurans (PCDFs) in surface freshwater systems in England and Wales. These chemicals are present naturally in the environment, but these natural sources are of relatively little importance compared with the emission of PCDD/Fs into the environment as unwanted by-products of anthropogenic processes.

The work involved the collection and analysis of surface waters and deposited sediments from a selection of rivers in England and Wales, including potentially contaminated and background sites.

The report showed very low levels of these dioxins in the water samples and confirmed that sediments are the main reservoir for PCDD/Fs in river systems as found elsewhere in Europe. There is little comparative data available for PCDD/F concentrations in UK rivers and no environmental standards, but the sediment results presented are comparable to the PCDD/F range found in UK soils and to surface river sediments in Europe.

Recommendations for the use of the results are given, including a sampling protocol for water and sediment sampling suitable for future NRA applications and for future PCDD/F monitoring studies which may be considered.

Key words: Dioxins, furans, PCDD, PCDF, sediments, water and rivers.

**NRA Publication Ref: R&D Project Record 330/4/A**

**Distribution of PCDDs and PCDFs in Surface Freshwater Systems**

Rose C L, McKay W A and Ambridge P F (1994)

AEA Technology

NRA Project No. 330

This report provides supplementary information to that found in R&D Note 242. The information presented includes:

- responses to the questionnaire sent to NRA Regions to identify both background sites and sites of potential PCDD/F pollution;

- detailed descriptions and maps of the sites sampled;

- the composition of the internal standard used for PCDD/F analysis of waters and sediments; and

- total concentration and toxic equivalent results from PCDD/F analysis of water and sediment samples.

Key words Dioxins, sediments, water, PCDD, PCDF, pollution.

**NRA Publication ref: R&D Note 183**

**Analysis, Storage and Archiving of Data**

Cremer and Warner Ltd (1993)

NRA Project No. 361

A review of practices in other industries and application areas revealed that, often, a more sophisticated approach to data management is adopted compared with current techniques used in the NRA. These techniques made fuller use of the data by correlation or comparison with other data and the use of rate of charge alarms and two or more levels of alarm threshold. Alternatively, in the process industry, a voting system is often used, in which three probes monitor the same parameter. This allows cross checking between probes for the ready detection of a fault probe and a reduction in the loss of data. Techniques used in other application areas are rule- and knowledge-based methods, pattern recognition techniques and neural systems. Analysis of the data exploited statistical packages, mathematical tool boxes and Geographical Information Systems.

A number of techniques were tested on automatically monitored water quality data. Initially, exploratory data analysis was used to gain an insight into the nature and quality of data.

Key words: Archiving, automatic, monitoring, pollution, rivers, water quality.

NRA Publication ref: R&D Project Record 361/4/NW provides further details in support of R&D Note 183

**NRA Publication Ref: R&D Note 121**

**Combined Distribution Method in Estuaries**

Evans G D, Ellis J C and Clark K J (1993)

WRc

NRA Project No. 337

This report describes the development and implementation of a method for simulating the long-term behaviour of the water quality of estuaries. The method is useful for management and control of estuary quality and, in particular, to assist discussions of applications for consents to discharge effluent to estuaries.

Examples are provided of application of the method to two UK estuaries for which one-dimensional models had previously been built, and descriptions are given of the new programmes and subroutines which have been written. In addition, a programme has been developed to simulate the results which would be obtained from the application of different sampling strategies to the output from the model.

Key words: Combined, distribution, method, estuary, water, quality, modelling, consents, discharges, statistics, programmes, time series, sampling, strategy.

### 3.5.3 Analytical Techniques

**Topic Leader:**
Dave Britnell, Principal Scientist with Thames Region.

**Rationale:**
The prime aim of the R&D in this Topic Area is to improve the efficiency and effectiveness of chemical analyses in the NRA. This is achieved by developing new and improved methodologies which maximise throughput and provide better limits of detection. This work will be an important part of market testing as it will help establish NRA preferred methods of analysis.

**Table 3.3 Analytical Techniques (Topic A3) R&D Programme (1994/95)**

| Proposal/Project No. | Title | Start/End | Contractor | Comments |
|---|---|---|---|---|
| | **On-going projects A3** | | | |
| A06(91)4 319 | Mothproofing agents and water quality management<br>To develop analytical methods for and examine the fate of mothproofing pesticides in the water environment and to assess the significance of the materials when discharged to aquatic systems. | 9/91<br>9/94 | University of Salford | |
| 062/035 | Microbiological techniques<br>To produce and update a manual of standard microbiological techniques for the NRA - which would include sampling analysis and quality control procedures.<br>Phase 3 - further techniques | 10/92<br>5/94<br><br>95/96<br>95/96 | WRc | Part-funded by SNIFFER |
| A09(92)8 460 | Use of supercritical carbon dioxide for extraction of trace organics<br>To develop an automated method for the selective extraction of Red List organics (except organotin and dioxin compounds) at levels useful for Red List monitoring work. | 3/93<br>3/95 | University of Leeds Innovations | |
| A03(93)1 527 | Farmstat pesticides<br>To develop analytical methods based on solid phase HPLC extraction and HPLC/MS to cover the increasing range of agrochemicals used in the environment. | 6/94<br>1/96 | In-house | |
| | **Proposed new starts A3** | | | |
| A09(92)3 | ELISAS for the screening of herbicides and pesticides<br>To establish a strategy for the introduction of immunoassays into the laboratories for the cost effective analysis of samples and for screening and semi-quantification of pesticides and herbicides | 94/95<br>94/95 | | |
| A03(94)2 | Improved techniques for dangerous substances in saline waters<br>To develop improved techniques to monitor dangerous substances in saline waters. | 94/95<br>94/95 | In-house | |
| A03(94)3 | New techniques for pesticides<br>To develop novel techniques for the analysis of pesticides to assist in pollution control activities. | 94/95<br>95/96 | WRc | |

## Outputs

No outputs were produced through this Topic Programme in 1993/94.

### 3.5.4. Instrumentation and Field Techniques

**Topic Leader:**
Paul Williams, Automatic Monitoring Officer with South Western Region.

**Rationale:**
The basis for research into new water quality instrumentation is two-fold; Firstly the appraisal of equipment currently available and secondly, the subsequent development of new items where they are needed operationally. The instrumentation needs to be accurate, reliable and robust, capable of measuring determinands used to assess the quality of surface waters. The NRA's National Centre on Instrumentation and Coastal Surveillance coordinates the R&D and operational aspects of such work.

**Table 3.4 Instrumentation and Field Techniques (Topic A4) R&D Programme (1994/95)**

| Proposal/ Project No. | Title | Start/ End | Contractor | Comments |
|---|---|---|---|---|
| | **On-going projects A4** | | | |
| A05(91)2 348 | Field detection of algal toxins<br>To develop a field test kit for the detection to a specified level of Microcystin - LR in water. | 9/91 5/94 | Biocode | Link with Project 349 |
| A05(91)4 349 | Validation of field procedures for algal toxin field test kits<br>To validate for NRA the development and performance of the field test kit for Microcystin - LR, developed by Biocode and develop field procedures for its use by NRA staff. | 9/91 5/93 | University of Dundee | Link with Project 348 |
| A10(90)4 240 | Bioaccumulation of Red List organic compounds<br>To develop code(s) of practice on the use of bioaccumulation techniques for monitoring Red List trace organic substances in freshwaters and estuaries that can be used through all the NRA Regions. | 9/90 5/93 | Northern Environmental Consultants | |
| A15(90)6 247 | Broad spectrum sensors<br>To develop and test in the laboratory and in the field a prototype instrument, incorporating biosensors, allowing a rapid assessment of toxicity of aqueous samples.<br>Phase 2 - further prototype development | 12/90 11/93<br><br>94/95 95/96 | WRc, Luton College | NRA contribution to DTi link funded project |
| A10(91)7 442 | Development of new techniques for the monitoring of ammonia in water, Phase 2<br>To develop and test in the laboratory and the field new and improved techniques for monitoring ammonia in water. | 1/93 9/94 | M Squared Technology | |
| A10(92)1 427 | Assessment of field monitors for consent monitoring<br>To assess accuracy, reliability, applicability and cost effectiveness of available equipment to enable the NRA to monitor alternative determinands to BOD and suspended solids.<br>Phase 2 - evaluation of the best monitors | 9/92 2/94<br><br>94/95 95/96 | WRc | Part-funded by SNIFFER |
| A10(92)2 473 | Review of field test kits<br>To review the use of field test kits in water quality monitoring to establish the feasibility and benefits of undertaking research to develop further test kits. | 5/93 9/93 | R Bogue & Partners | Liaison with Laboratory of Government Chemist and AEA |
| A04(93)2 507 | Evaluation of mini metal sensors<br>To evaluate bench-top metals analyser and to develop methods to enable environmental concentration in seawater of major saline metals to be determined *in situ* on board a survey vessel.<br>Phase 2 - production of prototype | 12/93 7/94<br><br>94/95 95/96 | Ecossensors | |
| A04(93)1 523 | Moored marine water quality monitor<br>To develop a prototype self-contained monitoring buoy in order to fulfil our statutory obligations to both monitor out to the three mile limit and to investigate the effect of polluting discharges. | 3/94 7/95 | In-House | |

| Proposal/ Project No. | Title | Start/ End | Contractor | Comments |
|---|---|---|---|---|
| A04(93)3 521 | Feasibility study of track analysis particles
To carry out a feasibility study to determine whether the CR-39 technique could be used in order to provide accurate, cost effective data for the NRA. | 2/94 3/94 | Phoenix Research Laboratory | |
| | **Proposed new starts A4** | | | |
| A04(94)1 | Biochemical oxygen demand predictor and hand held instrument
To develop a method of indicating the level of BOD in a discharge of receiving water to enable NRA staff to make on the spot decisions for further sampling or action.
Phase 1 - feasibility study | 94/95 94/95 | | |
| A04(94)2 | Oil in water - a review of existing monitors
To review current equipment used to monitor the presence and levels of oil in freshwaters to provide NRA staff with accurate assessment of their performance. | 94/95 94/95 | | |
| A04(94)4 | Inland use of airborne remote sensing
To review the algorithms for interpreting data from airborne remote sensing to monitor the quality of inland waters. | 94/95 94/95 | | |
| A04(94)5 | Instrumentation for self-monitoring
To specify and/or develop the equipment for self-monitoring and auditing of compliance to provide information for NRA staff through a business study. | 94/95 94/95 | | |

## Outputs

This section lists the principal outputs that have been produced through the Topic Programme in 1993/94.

R&D Note 246                   The use of *in situ* assays to assess river quality

## 3.5.5 Biological Assessment

**Topic Leader:**
Roger Sweeting, Regional Scientist with Thames Region.

**Rationale:**
The protection of the aquatic biology is one of the prime functions of Water Quality. In accordance with this, the R&D in the Biology Topic Area provides information on aquatic flora and fauna in order to support the development of biological methods for assessing and monitoring water quality. Research in this area provides important assistance to the establishment of SWQOs.

### Table 3.5 Biological Assessment (Topic A5) R&D Programme (1994/95)

| Proposal/ Project No. | Title | Start/ End | Contractor | Comments |
|---|---|---|---|---|
| | **On-going projects A5** | | | |
| A13(90)1 243 | Testing and further development of RIVPACS, Phase 2
To provide a more uniform, objective, user-friendly approach to the assessment of biological water quality for rivers within the UK. | 8/92 6/95 | IFE | |
| A13(90)3 242 | Faunal richness of headwater streams
To assess the conservation value of headwater stream macroinvertebrates and their contribution to catchment macroinvertebrate richness, and determine agricultural impacts upon them and propose a conservation strategy. | 10/90 1/95 | IFE | |

| Proposal/ Project No. | Title | Start/ End | Contractor | Comments |
|---|---|---|---|---|
| A08(91)3 354 | **Effects of low level contaminants on marine and estuarine benthic communities** To evaluate experimentally the effects of low levels of contaminants on marine and estuarine benthic communities. | 10/91 6/94 | PML | Contribution to DoE Water Directorate project |
| A12(91)4 396 | **Sediment toxicity test development - insoluble substances** To develop internationally standardised toxicity tests for use with sediments contaminated with sparingly water-soluble substances. | 2/92 3/94 | WRc | Part-funded by CEC STEP programme |
| A12(92)3 420 | **Methods manual** To establish standard operating procedures for ecotoxicological techniques and to produce and update a methods manual. | 9/92 8/95 | WRc | Part-funded by SNIFFER |
| A12(92)2 494 | **Method development** To provide suitable selection tests for the ecotoxicological assessment of effluent and receiving water quality. | 10/93 3/94 | WRc | |
| A(93)8 504 | **Biological assessment methods** To quantify and, where possible, control sources of variability in freshwater macroinvertebrate data for a range of river types and biological quality bands in order to increase the value of NRA data in water quality management. | 11/93 2/95 | WRc/IFE | |
| | **Proposed new starts A5** | | | |
| A05(94)1 | **Applications of artificial intelligence in river quality surveys** To undertake analysis of data held on the National Biological Database and to further investigate the distribution of macroinvertebrates throughout England and Wales. | 94/95 95/96 | | |
| A05(94)2 | **Biological techniques of still water quality assessment** To review and examine techniques and assemblage analysis for lakes, canals and ponds; leading to classification. | 94/95 96/97 | | |
| A05(94)3 | **Alternative methods of biological classification of rivers** To reassess work on macroinvertebrates and other biotic indicators in a European context. | 94/95 95/96 | | Links to Thames OI/CEN work |
| A05(94)4 | **Ecotoxicological Quality Assessment procedures** To develop quality assessment procedures for ecotoxicological methods and to provide accurate management information. | 94/95 94/95 | | Contribution to SNIFFER project |

## Outputs

This section lists the principal outputs that have been produced through the Topic Programme in 1993/94.

| | |
|---|---|
| R&D Note 82 | Toxicity of common pollutants to freshwater aquatic life - A review of ammonia, arsenic, cadmium, chromium, copper, cyanide, nickel, phenol and zinc on endigenous species |
| R&D Note 83 | Predicting the effects of ammonia to freshwater fish |
| R&D Project Record 023/12/T | Ecotoxicological effects |
| R&D Note 170 | The development of *Gammarus Pulex* feeding rate bioassay |
| R&D Note 171 | The development of the biochemical Glutathione-S-Transferase (GST) assay |
| R&D Note 172 | The development of estuarine toxicity test using an indigenous mysid |
| R&D Note 173 | A preliminary evaluation of the shell valve activity monitor |
| R&D Project Record 061/6/T | Biological methods for water quality assessment |
| R&D Note 198 | Statistical analysis of relationships between chemical and biological river quality data |

### 3.5.6 Consenting and Discharge Impact

**Topic Leader:**
Gerard Morris, Water Resources and Quality Manager with Northumbria & Yorkshire Region.

**Rationale:**
The control of pollution from point sources is a significant activity for the water quality function. The R&D in this Topic Area underpins all aspects of discharge control and assists in the development of consenting policy, evaluating the impact of new treatment technologies and the other methods of reducing wastewater.

**Table 3.6 Consenting and Discharge Impact (Topic A6) R&D Programme (1994/95)**

| Proposal/ Project No. | Title | Start/ End | Contractor | Comments |
|---|---|---|---|---|
| | **On-going projects A6** | | | |
| A01(90)4 305 | Nitrification rates in rivers and estuaries<br>To define by laboratory studies the influence on nitrification of temperature, ammonia concentration, dissolved oxygen, light, suspended solids and salinity in rivers and estuaries. | 4/91<br>1/94 | WRc | Part-funded by SNIFFER |
| A04(90)6 339 | Treatment process for ferruginous discharges from disused coal workings<br>To investigate practical low cost (with respect to operation and maintenance) processes for the treatment of iron contaminated discharges from abandoned coal workings, including pilot plant trials. | 10/91<br>12/93 | Imperial College | Part-funded by BOC Foundation |
| A01(92)1 432 | Assessment and reporting of the impact of CSO discharges on receiving waters<br>To develop an objective assessment procedure to enable the impact of episodic events on receiving waters to be reported, within the framework of the regulatory duties of the NRA. | 12/92<br>3/94 | WRc | Contribution to joint NRA/FWR project under UPM programme |
| A01(92)3 408 | Urban Pollution Management application methodology<br>To develop expertise and provide guidance for the application of the modelling tools that have been developed under the Urban Pollution Management programme. | 4/92<br>4/93 | WRc | Contribution to joint NRA/FWR project under UPM programme |
| A01(92)7 464 | UPM Management manual<br>To produce a manual which will allow practitioners on both the regulator and sewerage undertaker sides of the industry to implement the Urban Pollution Management methodology to assess the impact of storm sewer overflows in rivers. | 2/93<br>9/94 | WRc | Contribution to joint NRA/FWR project under UPM programme |
| A01(92)11 436 | Removal of coloured effluents from dyehouses<br>To research novel and cost effective processes for removing colour from dyehouse waste in order to minimise adverse effects on the water environment. | 1/93<br>10/95 | University of Leeds | Contribution to a BTTG project |
| A01(92)9 430 | National consent translation project<br>To establish a sound methodology for the neutral translation of consents for sewage treatment works' effluents into the format required by the EC Directive on Urban Waste Water Treatment.<br>Phase 2 - standard sampler specifications | 11/92<br>6/94<br><br>94/95<br>94/95 | FWR | Collaboration with water plcs, DoE Northern Ireland and SNIFFER |
| A07(92)13 456 | Efficacy and effects of wastewater disinfection<br>To undertake desk, laboratory and field studies of the efficiency and environmental effects of candidate disinfection processes to support NRA policies on consenting disinfected discharges.<br>Phase 4 - further discharge types and techniques | 2/93<br>6/94<br><br>94/95<br>95/96 | WRc/CES | |
| A(93)7 468 | Cost benefit assessment<br>To develop an economic benefit methodology for evaluating environmental benefits resulting from changes in water quality stemming from improvements in effluent quality. | 7/93<br>3/94<br><br>94/95<br>94/95 | WRc, via FWR | Part-funded by Water Service plcs, WSA & OFWAT |

| Proposal/ Project No. | Title | Start/ End | Contractor | Comments |
|---|---|---|---|---|
| A06(94)6 493 | **Toxicity based consents** <br> To further develop toxicity criteria in regulatory control, to test the protocols and procedures necessary for the application of toxicity based consents. | 10/93 3/96 | WRc | Part-funded by HMIP and SNIFFER |
| A07(91)3 485 | **Survival of particular viruses in seawater** <br> To assess the extent to which particular viruses can survive in the aquatic environment after discharge in body fluids to sewage in seawater <br> Phase 2 - practical assessments | 8/93 9/93 <br><br><br> 94/95 94/95 | St Bartholomews | |
| A07(93)1 490 | **Identification of oestrogenic substances in STW effluent** <br> To identify and quantify the component(s) present in sewage effluents that are responsible for the vitellogenic response in fish. | 9/93 3/96 | MAFF/Brunel University | Part-funded by MAFF |
| | **Proposed new starts A6** | | | |
| A06(94)1 | **Expert system for consenting discharges** <br> To develop an expert system to assist in the setting of discharge consents: Phase 1 - definition study. <br> Phase 2 - development of prototype | 94/95 94/95 <br> 95/96 95/96 | | |
| A06(94)2 | **UPM implementation project** <br> To demonstrate the implementation of the Urban Pollution Management Manual and other products to assist NRA staff in implementing the procedures. | 94/95 95/96 | WRc, via FWR | Part-funded by WSA |
| A06(94)3 | **Fate of detergents and associated chemicals in wastewater and rivers** <br> To assess the sources, transport and fate of detergents and similar chemicals to enable receiving waters to be protected. | 94/95 95/96 | | Part-funded by SDIA and WSA |
| A06(94)5 | **Organic deposition and benthic effects** <br> To produce a comparative predictive model relating the emission and settlement of suspended solids (organic carbon) to changes in the sediment dwelling microfaunal communities. | 94/95 95/96 | | Part-funded by Forth RPB, WSA and SNIFFER |
| A06(94)6 | **Implications of real time control on regulation of CSOs** <br> To undertake a scoping study to review the issues and identify the options for funding and developing knowledge of real time control on the regulation of CSOs. | 94/95 94/95 | | |

## Outputs

This section lists the principal outputs that have been produced through the Topic Programme in 1993/94.

R&D Note 48　　　　　　　　　　Mixing zone case study

R&D Note 210　　　　　　　　　Definition study for the implementation of toxicity-based consents

## Summary Information

The following summary is provided for the document available to the general public.

**NRA Publication Ref: R&D Note 48**

**Mixing Zone Case Study**

Nixon S C *et al.* (1993)

WRc

NRA Project No. 224

The control of industrial discharges to fresh and saline waters requires the setting of appropriate consent conditions. In the UK this is done by applying the Environmental Quality Objective/Environmental Quality Standard (EQO/EQS) approach and, in the case of discharges into tidal waters, by the designation of a "mixing zone" within which it is recognised that the EQS may be exceeded.

This project tested, by field surveys, the efficacy of a mathematical model in defining the mixing zone around the Tioxide Ltd outfall on the Humber Estuary.

This study has demonstrated that the depth-averaged concentration was not the most appropriate statistic on which the mixing zone should be judged, because at this site the highest tracer concentrations were found near the bed. If the zone to be protected includes the water near the bed, this should be considered when estimating the size of mixing zone to be expected from the design of the discharge hardware. Even though the case studies had limited success in validating the model for the Tioxide discharge, the employed method potentially offers the best opportunity for validating the mixing zone around other discharges. For this strategy to work successfully the boat would have to be moored within the plume. This would be aided by the use of a tracer which could be measured *in situ* or by the measurement, *in situ*, of a component of the discharge.

Key words: Mixing zone, tracer, models, EQS, water quality, coastal waters.

## 3.5.7 Rural Land Use

**Topic Leader:**
Bob Huggins, Environmental Manager with South Western Region.

**Rationale:**
The research in this area is targeted at studies in to the effects of rural land use generally on water quality. The R&D supports the activities of the Rural Land Use Group in influencing and developing methods to ameliorate and reduce pollution from agriculture and forestry sources.

**Table 3.7 Rural Land Use (Topic A7) R&D Programme (1994/95)**

| Proposal/ Project No. | Title | Start/ End | Contractor | Comments |
|---|---|---|---|---|
| | **On-going projects A7** | | | |
| A10(90)2 450 | Impact of pesticides on river ecology<br>To assess the impact of different pesticides on the structure and functioning of riverine ecosystems.<br>Phase 1 - literature review and definition study | 3/93<br>11/93 | WRc | |
| A07(91)4 424 | Occurrence of *Cryptosporidium*<br>To ascertain whether or not a link can be established in agricultural areas between livestock farm pollution incidents or farm husbandry and the occurrence of *Cryptosporidium* oocysts in samples of surface water. | 9/92<br>6/94<br><br>94/95<br>94/95 | WRc | |
| A02(90)3 205 | Management of acid lakes by regulating nutrients - Phase 2<br>To obtain a viable and effective alternative methodology of habitat amelioration to liming for soft water upland lakes by addition of phosphate and monitoring changes in ecosystem and buffering capacity in order to produce a management tool. | 4/93<br>3/96 | IFE | Part-funded by IFE, DoE, National Power, Powergen and Albright & Wilson |
| A02(90)1 230 | Measures for protecting upland water quality<br>To develop management practices required for practical implementation of Forest and Water Guidelines, in particular the optimisation of buffer strip width in forest planting. | 4/90<br>6/93 | WRc | Part-funded by SNIFFER and Forestry Authority |

| Proposal/ Project No. | Title | Start/ End | Contractor | Comments |
|---|---|---|---|---|
| A02(90)4 270 | Assessing the impact of forest clear-felling on stream invertebrates<br>To determine the consequences of large scale clear-felling on upland stream communities. | 2/91 2/94 | University of Aberystwyth | |
| A02(91)5 314 | Acid waters: Llyn Brianne project<br>To investigate the impact of liming treatments and land use change on streams, and to refine models for predicting deposition impacts and land use changes in the aquatic environment. | 4/91 3/94 | In-house | Part OI/part R&D Part-funded by DoE Air Quality Division |
| A02(91)4 368 | Impact of erosion of forest roads on water quality<br>To study the natural erosion processes and rates from mixed aggregate built roads in upland forests and the impacts of heavy vehicles used for timber extraction, in order to identify impact on water quality. | 11/91 9/94 | IH | Part-funded by DoE, Scottish Office and NERC |
| 119 | Total impact assessment of pollutants in rivers, Phase 3<br>To investigate the pollution of streams draining agricultural catchments and specifically, to develop a simple model of the movement of pesticides from the point of application to streams. | 4/92 5/94 | IH | Part-funded by ADAS and AFRC |
| 122 | Effects of agricultural erosion on watercourses<br>To identify the factors affecting the yield of high sediment yielding catchments, and to produce recommendations to reduce adverse impacts of soil erosion. | 1/90 8/93 | IH | |
| A03(91)2 359 | Optimum application rates for low rate irrigation<br>To identify the appropriate rate, time and frequency of application of dilute farm effluent to different types of land without causing water pollution. | 11/91 3/94 | SSLRC | Part-funded by ADAS and AFRC |
| A02(92)3 465 | Impacts of fine particulate outputs associated with timber harvesting<br>To quantify the impacts of timber felling upon stream particulate loads and to investigate methods to ameliorate long and short term sediment pollution associated with forestry land use and practice. | 7/93 7/96 | IH | Part-funded by NERC |
| A02(92)5 416 | Nitrogen module for the IH acidification model (MAGIC)<br>To develop a nitrogen process module for MAGIC and to apply the model to investigate increasing nitrate in atmospheric deposition. | 8/92 7/95 | IH | |
| A03(92)2 453 | Land management techniques<br>To develop management techniques for soil and nutrient conservation including the use of buffer zones and farm management plans.<br>Phase 1 - definition study<br>Phase 2 - development of models and practical techniques | 2/93 8/93<br><br><br>94/95 96/97 | WRc | |
| A03(92)3 434 | Pathogens from farming practices<br>To study the incidence of pathogenic bacteria and viruses in soils receiving livestock waste and to investigate the persistence and mechanisms of transport of these organisms in the soil. | 10/92 10/95 | WRc | Part-funded by SNIFFER |
| 001/012 | Sources, impacts and detection of farm pollution<br>To develop biological methods for the detection of organic pollution from farms and for assessing the effectiveness of remedial action. | 4/90 7/94 | WRc | |
| A02(92)10 462 | Zinc from watercress farms<br>To produce watercress with improved resistance to the crook root fungus (*spongospora subterranea f. sp.nasturti*) and watercress yellow spot virus in order to reduce the need to treat watercress with zinc and consequent contamination of watercourses. | 3/93 10/96 | Horticulture Research International | Part-funded by Watercress Growers Associations |
| A07(93)2 502 | Impact of conifer harvesting<br>To assess the impacts of conifer harvesting and replanting on upland stream water quality with a view to identification of ameliorative management strategies and the development of a model and guidelines for environmental impact assessment. | 1/94 12/96 | IH/ITE | Part-funded by NERC. Part OI/part R&D |

| Proposal/ Project No. | Title | Start/ End | Contractor | Comments |
|---|---|---|---|---|
| | **Proposed new starts A7** | | | |
| A07(94)1 | Alternative farming methods - dairy<br>To develop new systems of milk production to reduce the impact on water quality, in particular, levels of phosphate. | 94/95<br>98/99 | ADAS | In parallel with LINK project including ADAS, Thames Water, UKASTA, British Sugar and Perry Foundation |
| A07(94)2 | Alternative farming methods - arable<br>To develop more environmentally-friendly methods of rotational arable farming to reduce the pesticide usage. | 94/95<br>96/97 | | |
| A07(94)4 | Pesticide disposal<br>To develop best practice guidelines to ensure that waste pesticides are effectively disposed of. | 94/95<br>94/95 | | |
| A07(94)5 | Guidance for assessing and controlling non-point sources of phosphorus<br>To develop guidance for the assessment of the contribution of non-point sources to the phosphorus budget in rivers to assist in pollution prevention. | 94/95<br>94/95 | | |
| A07(94)6 | Treatment technology for farm wastes<br>To investigate the potential for developing new technology to treat farm wastes for disposal to land. | 94/95<br>95/96 | Wisdom Agriculture | Part-funded by BOC Foundation and MAFF |
| A07(94)7 | Best Practice Manual<br>To develop a manual of best practice to help NRA staff advise farmers. | 94/95<br>95/96 | | |

## Outputs

This section lists the principal outputs that have been produced through the Topic Programme in 1993/94.

R&D Note 77 — Forestry impact on upland water quality

R&D Report 11 — The disposal of sheep dip waste
- Effects on water quality

R&D Note 137 — Maintenance of slurry tanks

R&D Note 156 — Review of water quality implications of conifer harvesting in the UK
- 1. Literature review and recommendations for research

R&D Note 159 — Review of water quality implications of conifer harvesting in the UK
- 2. Unpublished results from ITE clearfelling studies and management options

R&D Note 108 — Nitrate reduction for protection zones: The role of alternative farming systems

## Summary Information

The following summaries are provided for those documents available to the general public.

**NRA Publication Ref: R&D Report 11**

**The Disposal of Sheep Dip Waste: Effects on Water Quality**

Blackmore J and Clark L (1994)

WRc

NRA Project No. 208

The aims of this project were to investigate the impact of the disposal of sheep dip waste, principally on groundwater quality, but also with regard to disposal to shallow soils overlying impermeable bedrock.

A national survey of sheep farming and dipping practices was carried out. A number of catchment-based studies was carried out in Devon, Sussex, Northern Ireland and Scotland. A total of six dip disposal sites in these areas was investigated. Questionnaires were circulated to farmers in catchments in Devon and Sussex to determine background information on dipping and dip disposal. Alternative disposal methods for waste dip are reviewed.

Evidence was found of pollution of the unsaturated zone, and groundwater at sites on aquifers where dip was disposed of at high loading rates. Pollution of surface water and soil contamination is likely at sites on shallow soils and impermeable bedrock where disposal is to soakaway. However, surface water pollution may be limited by high dilution rates involved. Determining peak stream concentrations is difficult with routine chemical sampling; biological monitoring of macroinvertebrates is a useful tool when monitoring for dip pollution. Recommendations aimed at controlling and preventing the pollution of surface and groundwaters by sheep dip waste are given.

Key words: Sheep dip, groundwater pollution, freshwater pollution, organophosphorus pesticide, organochlorine pesticide, synthetic pyrethroid, soakaway.

**NRA Publication Ref: R&D Note 137**

**Maintenance of Slurry Tanks**

Newman P J, Lewis S E, Newman P J and Lewis S E (1993)

ADAS

NRA Project No. 268

This manual is based on the findings of a survey conducted by ADAS. The survey was carried out on a small number of vitreous enamelled steel slurry stores in order to determine both the extent/occurrence of corrosion in the tanks and the likely effect on the structural integrity and life expectancy.

Key words: Corrosion, slurry, polyurethane.

**NRA Publication Ref: R&D Note 108**

**Nitrate Reduction for Protection Zones: The Role of Alternative Farming Systems**

1992

Elm Farm Research Centre

NRA Project No. 109

Results are presented from field studies in which nitrate leaching was monitored at a number of organic farms. Based on the results of these studies the extent to which organic farming systems may contribute to or limit nitrate leaching is reviewed. Key features of the system tending to reduce the potential for leaching are: the prohibition of nitrogen fertilizer use; the requirement of legume-based rotations; an emphasis on careful management of manures; restrictions on quantities of livestock feed brought on to the farm; limitations on stocking rates. Key features which may increase the potential for nitrate leaching are the cultivation of legume-based pastures; the potential for pollution from manures.

Key words: Arable, cover crops, cultivation, farming practice, ley, nitrate, nitrate leaching, organic farming, rotations, water sampling.

## 3.5.8 Groundwater Pollution

**Topic Leader:**
Bob Harris, Regional Groundwater and Contaminated Land Officer with Severn-Trent Region.

**Rationale:**
The impact of waste disposal activities on adjacent surface and groundwater quality together with pollution from other sources pose real problems for the NRA. The R&D in this Topic Area is designed to provide the required methods for evaluating the problem of groundwater quality as well as assessing the range of techniques available for remediation.

### Table 3.8 Groundwater Pollution (Topic A8) R&D Programme (1994/95)

| Proposal/ Project No. | Title | Start/ End | Contractor | Comments |
|---|---|---|---|---|
| | **On-going projects A8** | | | |
| A14(91)1 381 | Pollution potential of contaminated sites - Phase 2. To assess groundwater pollution potential of contaminated sites and to relate these leaching tests to target levels for acceptability with respect to groundwater. | 94/95 95/96 | WRc | |
| A07(90)1 295 | Geochemical process modelling. To identify all relevant processes, geochemical and/or biochemical, which can apply to the subsurface environment and which can be quantified for the purpose of deriving predictive transport models for operations use. | 6/90 3/94 | BGS | |
| A14(91)3 380 | Development on contaminated land - Phase 3. To provide guidance to developers and regulatory bodies in the assessment, specification, supervision and achievement of effective and safe remediation of contaminated land using the most appropriate techniques. | 2/94 1/96 | Consortium, via CIRIA | Contribution to CIRIA project |
| A08(93)5 514 | Risk assessment methodology for landfills. To develop a risk assessment methodology for the assessment of landfill engineering plans. | 11/93 3/95 | Golders Associates UK Ltd | Contribution to DoE Waste Management project |
| A08(93)10 519 | Reliability of sewers in ESAs. To assess the impact of sewer leakages on groundwater quality and to identify factors which give rise to the problem and steps which can be taken to minimise the risks in order for the NRA to produce a strategy to minimise the risk of future groundwater pollution. | 2/94 11/94 | CIRIA | Contribution to CIRIA project |
| A08(93)11 513 | Long-term monitoring of non-containment landfills. To produce long term monitoring data for landfill gas and leachate for uncontained landfill sites on different aquifers in the UK, in order to provide the NRA and DoE with the technical background needed to develop waste management policy. | 1/94 3/95 | WRc | Contribution to DoE project |
| A06(94)4 528 | Fire water and chemical spillage retention systems. To develop detailed technical guidance on NRA/DoE requirements for the planning, design and construction of containment systems for the prevention of water pollution following industrial accidents. | 5/94 4/96 | CIRIA | Part-funded by DoE Toxic Substances Division |
| | **Proposed new starts A8** | | | |
| A14(92)6 | Remediation of groundwater pollution for organic solvents. To develop and test a decision tree to enable the NRA to respond rapidly and rationally to both short- and long-term problems concerning organic solvents. | 94/95 95/96 | | |
| A08(94)1 | Contaminated soils and shallow systems. To clarify the flow and transport of pollutants in soils to underpin the vulnerability classification within the GPP. | 94/95 95/96 | | |

| Proposal/ Project No. | Title | Start/ End | Contractor | Comments |
|---|---|---|---|---|
| A08(94)2 | **Effects of old landfill sites on groundwater quality** <br> To assess the extent of groundwater pollution potential of old domestic waste overlying major aquifers to assist in management decisions. | 94/95 <br> 96/97 | | |
| A08(94)3 | **Leachate recirculation** <br> To review the extent of current knowledge on the recirculation of leachate to assess the short-term risks of water pollution. | 94/95 <br> 95/96 | | |

## Outputs

This section lists the principal outputs that have been produced through the Topic Programme in 1993/94.

| | |
|---|---|
| R&D Note 301 | Leaching test for assessment of contaminated land: Interim NRA Guidance |
| R&D Note 181 | Pollution potential of contaminated sites |
| R&D Project Record 381/6/T | Pollution potential of contaminated sites (A test methodology and assessment procedure based on leachability) |

## Summary Information

The following summaries are provided for those documents available to the general public.

**NRA Publication Ref: R&D Note 301**

**Leaching Test for Assessment of Contaminated Land: Interim NRA Guidance**

Lewin K *et al.* (1993)

WRc and NRA in-house

NRA Project No. 381

The objective of this study was to review published leaching test methods in order to identify an appropriate test to determine the leachability of materials from contaminated sites. Assessment was made on the ability of tests to comply with NRA requirements, these being that the resulting method should be appropriate to the purpose (simulating natural rainfall conditions), simple, inexpensive and capable of rapid turnaround. The resulting information will be useful in assessing the potential of the site to contaminate surface and groundwater.

Documentation resulting from earlier phases of the work was circulated for comment in May 1993. In response to these comments, and in the light of recent developments in Europe on leach testing for waste disposal purposes, a method is presented. This method is on an interim basis, for the purposes of general assessment of mainly inorganic determinands; additional, more complex test procedures may be required for specific situations and contaminants.

The recommended leach test involves agitating solid material with an unbuffered leachant for one 24-hour period, and replaces previous methods under discussion, as outlined in R&D Note 181 (Pollution Potential of Contaminated Sites: A Review, April 1993).

Key words: Contaminated land, groundwater and surface water pollution, leaching tests.

## 3.5.9 Pollution Prevention

**Topic Leader:**
Phil Chatfield, Principal Pollution Officer with Thames Region.

**Rationale:**
Preventing pollution from non-point sources is a key area for Water Quality activities. Such sources cover a wide range of areas from urban run off to waste minimisation. The R&D in this Topic Areas addresses such sources which are non-agricultural in origin.

**Table 3.9 Pollution Prevention (Topic A9) R&D Programme (1994/95)**

| Proposal/ Project No. | Title | Start/ End | Contractor | Comments |
|---|---|---|---|---|
| | **On-going projects A9** | | | |
| A01(91)9 410 | Review of pollution emergency arrangements and remedial measures<br>To review the NRA's response and effectiveness in dealing with pollution incidents in controlled waters, and so to specify identified areas in which improvements can be made to the current emergency service across the Regions. | 6/92<br>5/93<br><br>94/95<br>95/96 | WRc | |
| A08(90)4 271 | Production and fate of blue-green algal toxins<br>To investigate the production and fate of cyanobacterial toxins in the environment in order to elucidate effective management strategies related to cyanobacterial blooms and scums. | 12/90<br>8/94 | University of Dundee | |
| A10(90)5 206 | Environmental fate of persistent organic compounds<br>To investigate the occurrence of persistent organic compounds in fish flesh. | 4/90<br>3/94 | Liverpool Industrial Ecology Research Centre | |
| A01(92)10 425 | Control of pollution from highway drainage systems<br>To determine the impact of surface water discharges and to produce practical guidance on measures which can be taken to reduce pollution.<br>Phase 2 - new control measures | 10/92<br>12/93<br><br>94/95<br>95/96 | CIRIA | Part-funded by DTp. Contribution to CIRIA project |
| A17(90)2 292 | Development of a risk assessment tool for catchment control<br>To produce a pollution assessment tool, applicable on a site by site basis, to calculate the probability of that site causing unacceptable pollution to Controlled Waters. | 2/91<br>7/93<br><br>94/95<br>95/96 | SRD | Part-funded by AEA Safety & Reliability Directorate |
| A07(93)3 487 | Use of recombinant M13 bacteriophage for pollution tracing<br>To produce a method for tracing pollution sources with a readily available non-toxic organism that can be identified in the laboratory. | 8/93<br>10/93 | University of Lancaster | |
| A01(92)8 391 | Demonstration project for industrial wastewater minimalisation<br>To identify, assess and demonstrate the benefits arising from the minimisation of industrial wastewater through systematic and strategic practice in association with other regulators.<br>Phase 2 - further demonstration areas | 4/92<br>5/94<br><br>94/95<br>96/97 | CEST | Contribution to CEST project, part-funded by HMIP, Yorkshire Water and BOC Foundation |
| | **Proposed new starts A9** | | | |
| A02(94)2 | Distribution and dynamics of blue-green algae<br>To assess the occurrence, variability and detection of blue-green algae and their toxins as part of a pollution prevention and monitoring strategy. | 94/95<br>95/96 | | |
| A09(94)1 | Use of industrial by-products in road pavement foundations<br>To examine the potential for contamination of surface or groundwater as the result of using industrial by-products (including reclaimed materials) in road construction. | 94/95<br>94/95 | | Contribution to CIRIA project, part-funded by DoT and DoE |

## Outputs

As a new Topic Area in 1994/95, no outputs were produced in 1993/94.

## 3.6  Commission A Budget

The majority of the issues addressed through the Commission A programme are resourced through Grant in Aid. However, where projects are linked to the control of discharges, then expenditure is recharged to the charges made for discharge consents. Figure 3.1 shows the level of both actual and planned expenditure in this Commission. There are also significant resources provided by other research commissioning bodies with mutual interests in water quality to projects within Commission A.

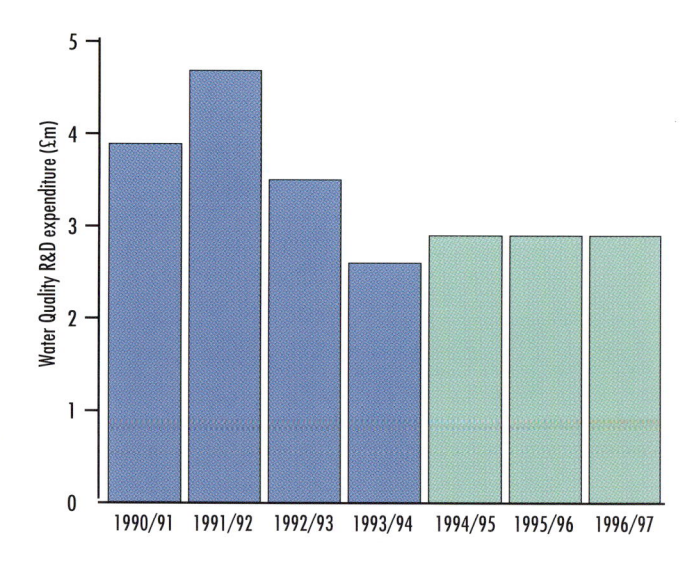

**Figure 3.1** Actual and planned expenditure on Water Quality R&D up to the financial year 1996/97

# 4. WATER RESOURCES

This section reviews the progress in Commission B R&D Programme over the year 1993/94 and sets out the planned Topic Programmes for 1994/95. The Water Resources-related R&D Programme is overseen by Richard Streeter, Water Resources Officer at Head Office, as Commissioner.

## 4.1 Business Rationale

The NRA's Water Resources Strategy, published in 1993, provides the business focus for much of the R&D undertaken in support of the Water Resources function. This Strategy sets out four principal activities, namely:

- Effective regulation;
- Monitoring;
- Operations; and
- An improved environment.

R&D is seen as a key tool in the NRA's ability to deliver these requirements in an efficient manner. In 1993/94, projects have supported functional activities such as the development of water resources strategy, demand studies and reviewing hydrometric practices.

R&D has also supported the NRA's efforts to alleviate low flows in certain parts of England and Wales as well as providing information essential in the establishment of river flow objectives.

## 4.2 Outputs Produced

A key output produced during 1993/94 was the computer model known as PHABSIM (Physical Habitat Simulation System). This model provides staff with the ability to predict the flow conditions required to support key aquatic species. In this respect, it will be an important tool in linking the problem of low flows with the setting of river flow objectives.

As part of effective regulation of water resources, the NRA has been keen to promote more efficient use of water by agriculture. With this in mind, the NRA published a report on the demand for irrigation water, providing a strategic view of the likely future needs of farmers.

A review of groundwater storage in British chalk aquifers was also produced in 1993/94. This work has assisted in the planning and development of water resources through providing standard information on these important resources.

## 4.3 Scientific Rationale

In mapping the water resources of England and Wales, the NRA is keen to ensure that decisions are based upon sound science. Significant advances have been made in the attribution of economic values to environmental aspects of rivers, a particularly important factor in the alleviation of low flows.

Habitat preference curves were developed for fish, invertebrates and macrophytes as part of the PHABSIM model. The development of these curves involved NRA staff snorkelling in rivers to observe the movement of trout and salmon in differing flows.

Work has continued on the development of a manual of aquifer properties. Jointly carried out by BGS, this project will produce a scientific reference of all aspects of aquifer in England and Wales and will contribute to the NRA's role in development control.

## 4.4 Strategic Projects

Four strategically important projects were started in 1993/94, the outputs for which will feed into key function initiatives. The determination of minimum acceptable flows will provide information to control the setting of abstraction licences and river flow objectives.

Further work on the costs and benefits of low flow alleviation has been started with the outputs from this R&D feeding into the NRA's phased implementation of low flow activities. Directly related to this is the provision of design tools for estimating low flows. The programme of work has been scoped, and the tools themselves will be developed in due course.

Finally, discussions have been held with BGS, DoE and FWR over the identification of the strategic issues involved in the management of groundwater. The scoping of these issues and programmes of work designed to address them will commence in 1994/95, and will be a significant project for the future development and protection of groundwater resources in the UK.

## 4.5 Topic Programmes and Outputs

This section covers the rationale behind the projects undertaken in each of the four Topic Areas within the Water Resources Commission. The outputs that have been produced between April 1993 and May 1994 are also included.

Summary information is provided for those outputs available to the general public; a complete listing of all outputs together with their availability is provided as an insert in the back cover.

Where an output is still under consideration by the NRA, where it contains commercial in confidence information, or where it feeds into another activity such as the development of a functional manual, its current availability may be less widespread. External organisations wishing to know more about such outputs should contact the relevant Topic Leader.

The R&D Programme for each Topic Area is provided in Tables 4.1 to 4.4. These tables list the projects either on-going or proposed for starting in the financial year 1994/95. Each Regional R&D Coordinator and the R&D Section at Head Office have further details on all such projects.

A fold-out guide to the information contained in the following sections is provided on the inside front cover.

### 4.5.1 Hydrometric Data

**Topic Leader:**
Geoff Burrows, Principal Resources Officer with Southern Region.

**Rationale:**
In line with the function strategic objectives, the prime direction of R&D in this Topic is to develop and improve hydrometric instrumentation and related techniques. The provision of useful data storage, retrieval and processing techniques will enable the function to obtain consistent, cost effective and reliable management information.

### Table 4.1 Hydrometric Data (Topic B1) R&D Programme (1994/95)

| Proposal/ Project No. | Title | Start/ End | Contractor | Comments |
|---|---|---|---|---|
| **On-going projects B1** | | | | |
| B01(92)2 478 | Calibration of portable electromagnetic current meters at low velocities. To evaluate the performance of all electromagnetic current meters currently available on the commercial market to ensure NRA staff are aware of the benefits and limitations of such meters. | 6/93 7/93 | S Walker/ HR Wallingford | |
| B01(93)2 529 | Current metering standards. To improve improve hydrometric instrumentation to provide more accurate data to the NRA. | 5/94 3/94 | HR Wallingford | |
| **Proposed new starts B1** | | | | |
| B01(92)4 | Enhancement of ultrasonic river flow gauges. To update and improve the measurement capability and accuracy of ultrasonic river flow gauges by the use of modern processes and electronic design together with improved software. | 94/95 95/96 | | |
| B01(93)1 | Performance assessment of ultrasonic and electromagnetic gauges at low velocities. To determine the performance of multipath ultrasonic and electromagnetic gauges at very low water velocities. | 94/95 95/96 | | Follows on from Project 478 |
| B01(94)1 | Evaluation of sewer flow monitors for abstraction measurement. To reach a decision on suitability of sewer flow monitoring equipment for monitoring abstractions and for discharges with low head, typically at fish farms and watercress beds. | 94/95 94/95 | | |
| B01(94)2 | Evaluation of acoustic doppler current profiler equipment. To identify the limitations of the equipment and to confirm that it operates within the manufacturer's specification to enable regional application. | 94/95 94/95 | | Links to Project 479 |
| B01(94)3 | Evaluation methodology for benefit of hydrometric networks. To develop the procedures for assessing (1) the requirements for hydrometric data, in an appropriate geographical area (2) the benefits derived from making such data available (3) whether or not the requirements are met by existing networks (4) what additions/reductions are necessary. | 94/95 95/96 | | |

### Outputs

No outputs were produced through this Topic Programme in 1993/94.

## 4.5.2. Flow Regimes

**Topic Leader:**
Mike Owen, Water Resources and Business Manager with Thames Region.

**Rationale:**
An effective appreciation of the inter-relationship between river flow and environmental factors is essential for achieving a balance between the availability of water resources and the impact upon the aquatic environment. The R&D in this Topic is focussed towards the issue and includes much of the work on low river flows.

### Table 4.2 Flow Regimes (Topic B2) R&D Programme (1994/95)

| Proposal/ Project No. | Title | Start/ End | Contractor | Comments |
|---|---|---|---|---|
| | **On-going projects B2** | | | |
| B02(91)3 316 | Effect of long term conifer afforestation and cropping in upland areas on water resources<br>To investigate the effects of forestry maturation and cropping on flow regimes and water chemistry of upland streams. | 4/91<br>3/94 | (a) IH<br>(b) Capital Works | Links with Project 114; part-funded by Forestry Authority and NWW plc |
| B2(93)3 491 | Design tools for low flow estimation<br>To scope the benefits and costs of a range of proposals by IH to define a programme of future collaborative research.<br>To improve and extend the application of enhanced MICROLOWFLOWS software by selected developments. | 9/93<br>5/94<br>94/95<br>95/96 | IH | Part-funded by NERC |
| B02(91)2 282 | Ecologically acceptable flows, Phase 2<br>To provide the framework for an objective method of evaluation of prescribed minimum flows based on the recognition of ecologically acceptable flows opposite to particular seasonal requirements of aquatic life forms. | 1/94<br>3/96 | IH/IFE | |
| B02(93)1 520 | Determination of minimum acceptable flows<br>To develop the concept of minimum acceptable flows (MAFs) and a policy for their application. | 2/94<br>12/95 | University of Birmingham | Links to Project 405 and 282 |
| B02(93)2 515 | Extended flow records at locations in England and Wales<br>To synthesize monthly records at 15 locations distributed throughout England and Wales for the period 1860-1922 to be used in the assessment of the yield of water resource systems.<br>Phase 2 | 6/94<br>6/94<br><br>94/95<br>95/96 | Climatic research Unit UEA | |
| | **Proposed new starts B2** | | | |
| B02(94)1 | Revision of SWK methodology for assessing low flows<br>To review and revise the new NRA methodology in terms of relevant criteria, associated data needs and scoring matrix to improve consistency of results between regions. | 94/95<br>94/95 | | |

## Outputs

This section lists the principal outputs that have been produced through the Topic Programme in 1993/94.

R&D Note 184  River bed lining
- State of the Art Review

R&D Note 185 and  Ecologically acceptable flows
R&D Project Record 282/1/Wx  - Assessment of instream flow incremental methodology

## Summary Information

The following summaries are provided for those documents available to the general public.

**NRA Publication Ref: R&D Note 184**

**River Bed Lining - State of the Art Review**

Ashby-Crane R *et al.* (1994)

Sir William Halcrow and Partners

NRA Project No. 419

The report defines the state of the art in river bed lining with respect to the available material sand methods, the environmental and engineering constraints and the methodology for project planning, design and implementation.

The results of the literature review indicate that there have been few additions relating to river bed lining since publication of the review by Watson Hawksley in 1990. Replies to the questionnaire reveal that only one project has been taken to the detailed design stage (River Slea). Experience in construction and monitoring of river bed lining schemes is limited to the Gussage Stream.

The available lining materials and methods have been examined and their suitability for particular site conditions assessed. The need for adequate surveys and consultations prior to construction, together with environmentally sensitive implementation and regular monitoring of the completed scheme is emphasised. A methodology is presented covering all aspects of implementation of river bed lining schemes, from design to post construction monitoring.

Key words: Channel lining, riverine environment, low flow rivers, groundwater abstraction.

**NRA Publication Ref: R&D Note 185**

**Ecologically Acceptable Flows**

Elliot C R N *et al.* (1993)

Institute of Hydrology

NRA Project No. 282

A national assessment by the NRA (1990) of low river flows identified 20 sites demanding urgent consideration. The current high profile of low flow conditions existing in UK rivers after two years of serious drought conditions, coupled with the requirement under the Water Act 1989 for the NRA to set Minimum Acceptable Flows when requested by the Secretary of State, has prompted the need to develop operational tools for managing aquatic communities in British rivers on a national scale.

The Instream Flow Incremental Methodology and Model PHABSIM were assessed by application on selected study reaches on ten different rivers in England and Wales. Study rivers were selected from ten different ecological groups identified by analysis of data from the RIVPACS database. At each of the study sites, hydraulic data have been collected at a number of calibration flows. Hydraulic models have been calibrated to simulate a wide range of discharges for nine of the relevant study sites. To assist in the choice of target species and in habitat suitability curve construction, the study sites have been electrofished on one occasion and invertebrates have been collected from selected microhabitats.

Habitat suitability curves have been constructed for three fish species (trout, roach and dace), two macrophyte species (Ranunculus and Nasturtium) and ten invertebrate species. Curves are based on expert opinion, existing data and information from the literature. These curves have been combined with hydraulic simulation outputs data to give habitat vs discharge relationships for selected target species. An example of the construction of habitat time series and habitat duration curves is given using data from the East Stoke Mill Stream.

Key words: Discharge, ecology, ecologically acceptable flows, habitat suitability curve, hydraulics, IFIM, PHABSIM, RIVPACS, velocity, weighted usable area.

### 4.5.3. Water Resources Management

**Topic Leader:**
Cameron Thomas, Resource Utilisation Engineer with Anglian Region.

**Rationale:**
The main accent of R&D in Topic B3 is to develop improved techniques for water resource assessment and management as a support to the progression of function strategies and policies in this area. The work is carried out on a catchment scale and on a national basis.

**Table 4.3 Water Resources Management (Topic B3) R&D Programme (1994/95)**

| Proposal/ Project No. | Title | Start/ End | Contractor | Comments |
|---|---|---|---|---|
| | **On-going projects B3** | | | |
| B03(92)5 484 | Evaluating the costs and benefits of low flow alleviation - Phase 2<br>To undertake surveys on two specified low flow rivers to obtain data for a cost benefit analysis in order for the NRA to justify alleviating low flows. | 7/93 7/94 | University of Middlesex | Follows on from Project 401 |
| B03(92)1 406 | Expert systems for water resources management - Phase 2<br>To provide an intelligent assistant computer program incorporating an expert system and data organiser to aid abstraction licence application determination and monitoring by NRA water resources officers. | 8/93 9/94 | University of Surrey | Follows on from Project 241 |
| B03(93)4 414 | Surface water yield assessment<br>To review and assess the suitability of existing methods for surface water yield assessment for both multiple and stand-alone sources and to recommend procedures for both sources for the NRA to adopt. | 7/92 10/93 | Wallace Evans | |
| B03(93)4 505 | Technical procedures for licence determination<br>To develop a methodology by which abstraction licence applications may be determined consistently and with due regard to protected rights and in-river needs. | 1/94 3/95 | Sir William Halcrow | |
| | **Proposed new starts B3** | | | |
| B03(93)1 | Groundwater resource reliable yield<br>To identify criteria and develop a methodology for estimating the yield of groundwater resources for use on a national basis to improve water resource planning. | 94/95 95/96 | | |
| B03(93)5 | Metering of water abstraction - good practice manuals<br>To produce "Good Practice Manuals" to be used as reference documents and a basis for training, giving the best type of metering for each application plus guidance on maintenance, calibration and accuracy. | 94/95 95/96 | | |
| B03(94)3 | Demand forecasting issues and methodology<br>To derive a forecasting methodology for Public Water Supply and industrial abstraction incorporating current initiatives within the water industry and enabling the economic level of demand management to be determined and to improve knowledge of present components of water demand. | 94/95 95/96 | | Part-funded by WSA |
| B03(94)4 | Incentive charges: Practical difficulties of measurement<br>To identify the "weights and measures" type of standards which would be required for quantity-based abstraction charges and to identify the practical problems and costs of meeting these standards by completing a field-based review for each type of abstraction including Public Water Supply, Spray Irrigation, Industry and Agriculture. | 94/95 95/96 | | |

## Outputs

This section lists the principal output that has been produced through the Topic Programme in 1993/94.

R&D Project Record 413/3/A          Demand for irrigation water

## Summary Information

The following summaries are provided for those documents available to the general public.

**NRA Publication Ref: R&D Project Record 413/3/A**

**Demand for Irrigation Water**

Weatherhead E K *et al.* (1993)

Silsoe College, Cranfield University

NRA Project No. 413

The work has provided predictions for the likely future demand for irrigation water in England and Wales and given recommendations on possible responses. The study was restricted to agricultural and horticultural irrigation and took no account of any long-term climate change.

The results take into account the drought years of 1989 - 1991 and eliminate any bias from overestimation due to this factor. The 'most likely' prediction for growth in volumetric demand is given as 1.7% per year from 1996 to 2001 and 1% per year from 2001 to 2021 for the 'dry' year.

The report makes recommendations for NRA responses.

Key words: Irrigation, water demand, water resources, predictions and agriculture.

### 4.5.4. Groundwater Protection

**Topic Leader:**
Mike Eggboro, Catchment Resources Manager with North West Region.

**Rationale:**
The research in this Topic Area supports the function initiative of resource protection. It aims to provide an understanding of groundwater processes and the ability to monitor and control, through aquifer protection policies, the quality of groundwaters both across England and Wales and in time.

**Table 4.4 Groundwater Protection (Topic B4) R&D Programme (1994/95)**

| Proposal/ Project No. | Title | Start/ End | Contractor | Comments |
|---|---|---|---|---|
| | **On-going projects B4** | | | |
| B04(91)1 306 | Bacterial denitrification in aquifers<br>To investigate the controls on bacterial denitrification in the unsaturated zone of the Chalk and Sherwood Sandstone aquifers in Nitrate Sensitive Areas. | 4/90<br>5/93 | BGS | |
| B04(93)1 499 | National system for recharge assessment<br>To develop an accurate, consistent and publicly defensible method of estimating mean annual groundwater recharge applicable to drift free areas of aquifer outcrop in England and Wales. | 10/93<br>10/94<br><br>94/95<br>95/96 | IH | Part-funded by NERC |
| B04(92)6 454 | Manual of aquifer properties<br>To assemble and publish and Aquifer Properties Manual for England and Wales in order to underpin water resources management, particularly the groundwater protection policy. | 3/93<br>3/96 | BGS | Part-funded by NERC |
| B04(92)5 455 | Hydrogeological characterisation of clays<br>To assess parameters relevant to superficial clay cover, identify mission data and pursue field evaluation of techniques for measuring these data. | 2/93<br>1/96 | BGS | Part-funded by NERC |

| Proposal/<br>Project No. | Title | Start/<br>End | Contractor | Comments |
|---|---|---|---|---|
| | **Proposed new starts B4** | | | |
| B04(93)2 | Strategic groundwater research review<br>To undertake a strategic review with BGS and other researchers/<br>customers into the priority areas of applied strategic research<br>which are of concern to the NRA and other customers, and into<br>more strategic issues relevant to the work of BGS and other<br>research groups. | 94/95<br>94/95<br>94/95<br>95/96 | BGS | Part-funded by NERC<br>and FWR |

## Outputs

This section lists the principal outputs that have been produced through the Topic Programme in 1993/94.

R&D Note 169 and  Groundwater storage in British aquifers
R&D Project Record 128/8/A  - Chalk

## Summary Information

The following summaries are provided for those documents available to the general public.

**NRA Publication Ref: R&D Note 169**

**Groundwater Storage in British Aquifers - Chalk**

Lewis M A *et al.* (1993)

British Geological Survey

NRA Project No. 128

The Chalk is the major aquifer in England, both in terms of area and quantity and quality of water abstracted from it, providing over 50% of the groundwater supplies in England.

Two catchments have been studied in order to improve our knowledge of the resources of the aquifer and its susceptibility to pollution. Rock volumes for the Chalk have also been calculated to allow the determination of the available stored water in the aquifer. The report gives details of the amount of water which is available from the aquifer under various conditions.

The study has provided the first estimates of the volume of groundwater stored in the Chalk aquifer and has highlighted the areas where our knowledge of aquifer parameters and groundwater movement are lacking. Key areas where further work is required have been identified.

Key words: Aquifer, chalk, delayed recharge, groundwater, porosity, specific storage, specific yield, storage coefficient and water resources.

## 4.4 Commission B Budget

The R&D in Commission B is resourced through revenue from abstraction licences and the budget reflects the level of information, techniques and new methods, required by the function to carry out its activities. Figure 4.1 shows the level of both actual and planned expenditure in this Commission. However, the NRA is not the only organisation with a need to carry out R&D in this areas. Part-funded work with the British Geological Survey, Institute of Hydrology and others ensures that the NRA maximises the use of its resources in this area.

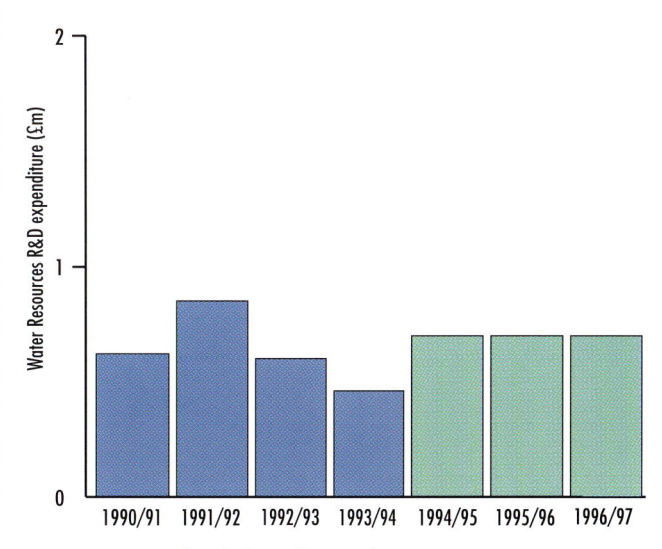

**Figure 4.1 Actual and planned expenditure on Water Resources R&D up to the financial year 1996/97**

# 5. FLOOD DEFENCE

This section sets out the progress made in the Commission C R&D Programme in 1993/94. Also described here, are the Topic Programmes for the 1994/95 financial year, together with the strategic direction of the R&D required in support of the Flood Defence function. This area of R&D is overseen by Lindsay Pickles, Flood Defence Officer at Head Office, as Commissioner.

## 5.1 Business Rationale

The NRA published its Flood Defence Strategy in 1993, setting out its objectives and key activities over a ten year horizon. The business needs for Flood Defence can be broadly placed in three categories, namely:

- Advice on the risk of flooding;
- Flood forecasting and warning; and
- Flood alleviation, through operational management and defence schemes.

The NRA is a statutory consultee, and advises Planning Authorities on areas at risk of flooding as part of the planning process. This advice is aimed at satisfying the requirements of Circular 30/92, published by DoE, on Development and Flood Risk. R&D has driven much of the work in this area, and will continue to support NRA activities.

The NRA has powers to alleviate existing flooding where economically beneficial. In order to achieve this, an objective identification and appraisal of works is required. In addition, effort has gone into ensuring that schemes are effectively prioritised, and it is here, through projects on asset management and coastal and river infrastructure management systems, that R&D has been used to support function priorities.

Flood defences can prove inadequate either because development has taken place in inappropriate areas or because floods are greater that expected. Where possible, the NRA provides flood warnings. A sound understanding of the techniques of flood forecasting and the problems associated with flood warning is an integral component of the R&D.

## 5.2 Outputs Produced

A key element of Flood Defence activities is the operational management of rivers. This is carried out with the aim of maintaining the carrying capacity of river channels. In early 1994, the NRA published a report summarising the R&D undertaken to improve the efficiency and effectiveness in Flood Defence operational management. R&D Report 7 covered all the major items of R&D undertaken between 1990 and 1993 within the Operational Management Topic Area.

A further project, designed to improve value for money in this area, has provided a mechanism for the economic appraisal of non-grant aided works. This output will assist staff in balancing the costs and benefits of this Flood Defence activity.

In early 1994 the NRA published, in association with MAFF, SERC (now EPSRC) and HR Wallingford, interim guidelines on the calculations required for the design of straight and meandering compound channels. R&D Report 13 is now being evaluated by design engineers both within the NRA and other organisations.

Guidance has also been produced on the design and operation of trash screens as a result of a project which completed in 1993. This Flood Defence publication (P-126) has provided essential information for NRA staff in ensuring that such screens are operated effectively.

## 5.3 Scientific Rationale

The Flood Defence function is strongly oriented towards operational tasks, with much of the information from R&D being used in day-to-day activities within the Regions. With this in mind, the rationale of the Commission C programme is to provide best operational practice in a form that allows a quick uptake of results.

The hand calculation methodology provided in R&D Report 13, described above, has substantially advanced this area of engineering technology. The new equations and algorithms will provide a benchmark for design engineers internationally.

In introducing new technical approaches to issues, reference often has to be made to experiences outside of the UK. The NRA has looked to the USA for experience on the stabilization of soil embankments, where widespread use of soil nailing has been made to improve stability.

## 5.4 Strategic Projects

Liaison with MAFF is seen as an important element in the development of strategic Flood Defence R&D projects. The general principle underpinning this area is that the NRA's R&D is focused towards more practical outputs whilst still underpinning the NRA's strategy in this area.

Two important strategic projects continue to address fundamental issues for coastal management. Firstly, a coordinated programme of R&D is designed to provide management tools for the management of saltmarshes as natural flood defences. This work is complemented by a similar study investigating the use and maintenance of beaches to protect coastal areas against flooding.

The NRA is also working closely with the Met Office to develop a thunderstorm and deep convection warning system in order to improve the detail of flood warnings it provides. This project is developing a system known as GANDOLF (Generating Advanced Nowcasts for Deployment in Operational Land-surface Flood Forecasting).

## 5.5 Topic Programmes and Outputs

This section covers the rationale behind the projects undertaken in each of the six Topic Areas within the Flood Defence Commission. The outputs that have been produced between April 1993 and May 1994 are also included.

Summary information is provided for those outputs available to the general public; a complete listing of all outputs together with their availability is provided as an insert in the back cover.

Where an output is still under consideration by the NRA, where it contains commercial in confidence information, or where it feeds into another activity such as the development of a functional manual, its current availability may be less widespread. External organisations wishing to know more about such outputs should contact the relevant Topic Leader.

The R&D Programme for each Topic Area is provided in Tables 5.1 to 5.6. These tables list the projects either on-going or proposed for starting in the financial year 1994/95. Each Regional R&D Coordinator and the R&D Section at Head Office have further details on all such projects.

A fold-out guide to the information contained in the following sections is provided on the inside front cover.

### 5.5.1 Fluvial Defences and Processes

**Topic Leader:**
David Wilkes, Tidal Defence Manager with Thames Region.

**Rationale:**
The R&D in this Topic Area will support the specialised field of engineering hydrology and hydraulics for the practising engineer. The outputs will enable better design and maintenance practices in the function as a whole. The work will also cover fluvial processes that may affect structure design.

**Table 5.1 Fluvial Defences and Processes (Topic C1) R&D Programme (1994/95)**

| Proposal/ Project No. | Title | Start/ End | Contractor | Comments |
|---|---|---|---|---|
| | **On-going projects C1** | | | |
| C01(90)1 252 | SERC flood channel facility, Phase 4<br>To provide reliable procedures for assessing the hydraulic performance and stage discharge function of two-stage or compound river channels taking into account sediment movement. | 5/94<br>6/94<br><br>94/95<br>96/97 | HR | Collaborative programme with EPSRC, MAFF and CEC |
| C01(91)2 333 | Infiltration methods for runoff control<br>To produce a manual of good practice for the design, installation and maintenance of the range of practical infiltration techniques currently available for the effective disposal of surface water drainage. | 8/91<br>8/93 | CIRIA | Contribution to CIRIA project also funded by DoE and Water Services plcs |
| C01(91)3 366 | Large-scale model investigation of a two-stage channel, Phase 2<br>To advance understanding of environmentally sensitive and cost-effective river channel design to enable appropriate flood defence standards of service to be provided. | 11/91<br>3/93 | University of Bristol | |
| C05(90)3 300 | Design and operation of trash screens<br>To establish best practice for the design and operation of trash screens, and to produce a manual of best practice.<br>Phase 3 - monitoring schemes | 10/91<br>9/93<br><br>94/95<br>96/97 | Posford Duvivier | |

| Proposal/<br>Project No. | Title | Start/<br>End | Contractor | Comments |
|---|---|---|---|---|
| C05(91)1<br>384 | Sediment and gravel bed transportation<br>To carry out field studies and monitoring of selected sites to improve management practice. | 2/92<br>5/94 | University of Newcastle upon Tyne | |
| C05(91)4<br>363 | Pumping stations - efficiency, operation and life-cycle costs<br>To review national practices and philosophy of pump specifications, configuration, telemetry, operating rules and energy management, to identify best practice. | 12/91<br>3/94 | Bullen & Partners | |
| C01(91)1<br>407 | Review of fluvial R&D related to flood defence<br>To identify and prioritise a strategic NRA research programme in the specific area of fluvial defences and processes and to ensure that the NRA's programme interfaces with other external and internal cost programmes. | 4/92<br>3/94 | Binnie & Partners | |
| C01(91)1<br>394 | Rainfall frequency studies, Phase 1b and 2<br>To review current methods used for rainfall frequency analysis and to develop new procedures where the current methods give unsatisfactory results. To compile the new procedures to form a volume of the proposed flood estimation handbook entitled "Rainfall Frequency Estimation". | 2/94<br>5/96<br>95/96<br>95/96 | IH | Memorandum of Understanding between Met Office, IH and NRA |
| C01(93)1<br>508 | Benchmarking for models<br>To produce standard benchmarks against which to test fluvial flood defence models in order to provide the NRA with a decision-making tool on appropriateness for different applications.<br>Phase 2 - benchmarking trials | 1/94<br>7/94<br><br>94/95<br>95/96 | Sir William Halcrow | |
| | **Proposed new starts C1** | | | |
| C01(94)1 | Monitoring methods and techniques for fluvial defences<br>To determine what fluvial parameters should be monitored and how to meet the needs of Flood Defence and integrate with other river management parameters. | 94/95<br>96/97 | | Links with proposal F01(94)1 |
| C01(94)2 | Scoping study and design notes for fluvial design manual<br>To provide a framework, format and cross-referencing system to draw together new knowledge and techniques in a fluvial design manual. | 94/95<br>96/97 | | |

## Outputs

This section lists the principal outputs that have been produced through the Topic Programme in 1993/94.

R&D Report 13            Design of straight and meandering compound channels
                         - Interim guidelines on hand calculation methodology

R&D Project Record 300/2/T       Design and operation of trash screens
and Flood Defence Publication P-126   - Interim guidance

## Summary information

The following summaries are provided for those documents available to the general public.

**NRA Publication Ref: R&D Report 13**

**Design of Straight and Meandering Compound Channels - Interim Guidelines on Hand Calculation Methodology**

Wark J B, James C S and Ackers P (1993)

HR Wallingford

NRA Project No. 252

A compound channel consists of a main channel, which accommodates normal flows, flanked on one or both sides by a flood plain which is inundated during high flows. For water levels above the flood plain, the flow is strongly influenced by the interaction between the fast-flowing water in the main channel and the relatively slow-moving water over the flood plains. This significantly complicates the estimation of stage-discharge relationships. The extra turbulence generated by the flow interaction introduces energy loss over and above that associated with boundary resistance. This is not accounted for by the conventional resistance equations and their direct applications may result in significant error.

This report brings together the two procedures for straight and meandering compound channels in one document, to facilitate their use in the design office. Chapter 2 summarises the important mechanisms which affect the discharge capacities of straight and meandering compound channels. Chapter 3 provides guidelines on the choice of the straight or meandering method. The details of the two procedures are given in Chapter 4, along with detailed worked examples. Annexes at the end of the report give summaries of the development and verification steps which were followed for the two procedures. Implications for software are also given.

Both the guidelines given in this report and the layout are regarded as interim. It is issued in the hope that users will provide feedback on the procedures and their use.

Key words: Hydraulic, design, compound channels, straight, meanders, flood plains, stage-discharge, bed shear stress, worked examples, National Rivers Authority, HR Wallingford.

**NRA Publication Ref: NRA Project Record 300/2/T**

**Design and Operation of Trash Screens**

1992

Posford Duvivier

NRA Project No. 300

This Phase reviewed current experience and knowledge on the subject of trash screens to enable interim guidance to be produced on best design practice and operation.

A methodology was established to examine and analyse the research previously undertaken on the subject, identifying deficiencies in the information which required further detailed investigation, and obtaining a legal view concerning responsibilities of trash screen sites.

It was concluded that major problems which exist at trash screen sites have changed little over the past five years with capacity and clearance difficulties causing most concern.

Experience on automatic mechanical screens was found to be limited and so far has produced as many problems as have been solved by their introduction. The report concludes that the installation of such screens in aggressive urban environments should be monitored closely until their performance is better understood. It was recognised that under certain conditions, the use of mechanical screens operated manually could provide an effective and efficient emergency response.

Key words: Automatic screens, culvert screens, debris screens, mechanical screens, telemetry, trash screens.

Flood Defence Publication P-126 provides operational guidance to NRA staff, consultants and others to aid design and operation of trash screens in water courses.

### 5.5.2 River Flood Forecasting

**Topic Leader:**
Bob Hatton, Flood Defence Manager with South Western Region.

**Rationale:**
Emergency response is an important issue for the NRA, an integral part of which is the role of flood forecasting. The primary direction of the R&D directed towards the issue is the development of flood forecasting systems, including the use of weather radar information, to enable accurate forecasts of river stage heights.

**Table 5.2 River Flood Forecasting (Topic C2) R&D Programme (1994/95)**

| Proposal/ Project No. | Title | Start/ End | Contractor | Comments |
|---|---|---|---|---|
| | **On-going projects C2** | | | |
| C02(90)3 287 | Evaluation of integrated flood forecasting models. To evaluate the integrated flood forecasting systems currently being commissioned or operational in the NRA, and to provide recommendations for future development. | 1/92 3/95 | C T Marshall | Follows on from Project 201 |
| C02(91)1 372 | Continuous monitoring of soil moisture for flood hydrology. To install and test in the field instrumentation to provide continuous measurements of soil moisture and to assess the value of measurements from such instruments in an operational flood forecasting system. | 1/92 10/94 | IH | |
| C02(90)4 367 | Use of vertically-pointed radar. To obtain vertical radar profiles for specific sites and specific meteorological events, in order to overcome key problems associated with the existing national network of radars. | 12/91 5/94 | University of Salford | |
| C02(91)2 357 | Development of fully distributed models using radar rainfall data. To develop the methodology to use radar data from 2 km and 5 km squares to construct distributed models of rainfall on a catchment basis. | 11/91 5/94 | IH | |
| C02(91)3 315 | Improved adaptive calibration technique for weather radars. To develop an improved adaptive calibration technique for weather radars which overcomes some existing shortcomings for hydrological forecasting. | 4/91 3/94 | University of Lancaster | |
| C02(92)1 470 | Improving the accuracy of radar rainfall data. To compare accuracy of different methods of weather radar precipitation measurement and data processing techniques. | 3/93 11/95 | All Water Technology | Contribution to HYREX project |
| C02(92)2 448 | Development of improved methods for snow melt forecasting. To develop improved models for assessing the water content of snow packs and forecasting snow melt. | 2/93 2/95 | IH | |
| C02(93)2 510 | Thunderstorm and deep convection warning system. To develop a system for alerting hydrologists to the likely occurrence of thunderstorms and for forecasting the heavy rainfall associated with thunderstorms and deep convection through development of the GANDOLF system (Generating Advanced Nowcasts for Deployment in Operational Land-surface Flood Forecasting). | 1/94 3/96 | Metstar | Part-funded by Met. Office |
| | **Proposed new starts C2** | | | |
| C02(94)1 | Comparison of calibration techniques to determine best approach. To compare radar rainfall calibration techniques and identify best practice. | 94/95 95/96 | | |

### Outputs

No outputs were produced through the Topic Programme in 1993/94.

## 5.5.3 Catchment Appraisal and Control

**Topic Leader:**
John Gardiner, Technical Planning Manager with Thames Region.

**Rationale:**
The role of enforcement and regulation for Flood Defence is an important one and is supported by R&D in Topic Area C3. It aims to develop new and existing approaches for assessing and influencing the impact and extent of developments within the floodplan.

**Table 5.3 Catchment Appraisal and Control (Topic C3) R&D Programme (1994/95)**

| Proposal/ Project No. | Title | Start/ End | Contractor | Comments |
|---|---|---|---|---|
| **On-going projects C3** | | | | |
| C03(91)1 426 | Forward planning process; Best European Practice  To study experiences in other EC member states and limited examples elsewhere with integrated flood plain land use and flood defence, taking into account the conservation and enhancement of the river environment. | 8/92 7/94 | Middlesex Univ. Flood Hazard Research Centre | |
| **Proposed new starts C3** | | | | |
| C03(93)1 | Economic and environmental appraisal  To develop economic tools for analysing the impact of catchment control in order to link economic appraisal to environmental assessment. | 94/95 96/97 | | |
| C03(93)2 | Analysis of catchment control problems  To analyse the failure in planning liaison to improve the development of the control in planning of Flood Defence activities. | 94/95 95/96 | | |
| C03(94)1 | Developing new strategic instructions and guidelines  To provide enhanced national guidance for influencing development in flood risk areas. | 94/95 95/96 | | |
| C03(94)2 | Developers contribution to works  To consider the economics of development in the flood plain and determine the extent of contributions from proposed developers for NRA schemes. | 94/95 95/96 | | |

## Outputs

This section lists the principal output that has been produced through the Topic Programme in 1993/94.

R&D Note 313 and
R&D Project Record 345/2/T

Public perception of rivers and flood defences
- Flooding and flood defences in York

### 5.5.4 Operational Management

**Topic Leader:**
Gary Lane, Regional Flood Defence Manager, Southern Region.

**Rationale:**
The development of a framework for the management of NRA flood defence maintenance is the central theme for this Topic Area. This will enable the production of consistent, prioritised and cost effective work programmes that also recognise the interest of other core functions.

**Table 5.4 Operational Management (Topic C4) R&D Programme (1994/95)**

| Proposal/ Project No. | Title | Start/ End | Contractor | Comments |
|---|---|---|---|---|
| | **On-going projects C4** | | | |
| C04(91)2 341 | Asset management planning for flood defence<br>To develop a manual which identifies the variety of flood defences in existence, their strengths, failings and relative vulnerability, the management of their maintenance and, where appropriate, the methods available for rehabilitation. | 8/91<br>10/94 | Binnie & Partners | |
| C04(91)1 317 | Evaluation of alternative river maintenance strategies<br>To develop and verify a database and methodology which assesses the benefits associated with alternative river maintenance strategies in rural, mainly agricultural catchments. | 4/91<br>6/94<br>94/95<br>95/96 | Silsoe College | |
| C04(93)1 488 | Aquatic weed control operation - Phase 2<br>To produce best practice guidelines for aquatic weed control throughout the NRA in order to promote efficient and effective management practices. | 2/94<br>6/95 | AWRU | |
| C04(93)2 512 | Evaluation of Aquatic Weeds Research Unit<br>To undertake an evaluation of the past work undertaken by the AWRU for its sponsors, and to recommend how their requirements for information and advice on the management of aquatic weeds should be provided in the future. | 1/94<br>3/95 | CEST | |
| C04(90)3 213 | Grass management operations, Phase 2<br>To assess routine mowing operations throughout the NRA; to implement further research identified in Phase 1: to carry out field trials; and to produce guidance note on best practice for riverbank grass management. | 94/95<br>95/96 | | |
| C04(92)3 435 | Economic appraisal of non-grant aided schemes<br>To develop a method for economic appraisal of flood defence works not covered by MAFF grant aid. | 1/93<br>3/94 | Mott MacDonald | |
| C04(92)1 516 | Management of shoaling/desilting operations<br>To review guidance on the safe, economic and effective disposal and management of dredgings in order to develop best practice guidelines.<br>Phase 2 | 2/94<br>10/94<br><br>94/95<br>94/95 | CIRIA | Liaison with Project 384 |
| | **Proposed new starts C4** | | | |
| C04(94)1 | Quality assurance for survey techniques<br>To determine what quality of survey work the NRA should specify. | 94/95<br>95/96 | | |

Annual R&D Review - 1994

## Outputs

This section lists the principal outputs that have been produced through the Topic Programme in 1993/94.

R&D Report 7                    Improving the efficiency and effectiveness in Flood Defence operational management
                                - Review of R&D (1990-1993)

R&D Note 189                    Aquatic weed control operations
                                - Existing Practice

R&D Note 187 and                Economic appraisal of non-grant aided work
R&D Project Record 435/2/NW

## Summary Information

The following summaries are provided for those documents available to the general public.

**NRA Publication Ref: R&D Report 7**

**Improving Efficiency and Effectiveness in Flood Defence Operational Management - Review of R&D (1990-1993)**

Mott MacDonald and Gould Consultants (1994)

NRA Project No. 373

The NRA spent over £230 million on Flood Defence capital and maintenance in 1992/93. This document describes the research and development undertaken over the period 1990 to 1993 to develop new procedures to ensure that Flood Defence operations are adequately assessed and prioritised and are undertaken to consistent standards in a cost-effective manner. The projects are presented within the rationale of the integrated framework which has now been developed for Flood Defence operational management. This document, explaining the underlying R&D, is therefore a useful precursor to the Flood Defence Management Manual to be published later.

Key words: Flood Defence, management, assets, standards of service, fluvial maintenance, grass management.

### 5.5.5 Coastal and Tidal Defences and Processes

**Topic Leader:**
Robert Runcie, Regional Flood Defence Manager with Anglian Region.

**Rationale:**
Uppermost of the priorities for this Topic Area is the development of cost effective and environmentally-sympathetic engineering options for coastal and estuarine flood defences. The R&D will also support the planning and execution of coastal management techniques as well as assisting the strategy to compensate for sea level rise.

**Table 5.5 Coastal and Tidal Defences and Processes (Topic C6) R&D Programme (1994/95)**

| Proposal/ Project No. | Title | Start/ End | Contractor | Comments |
|---|---|---|---|---|
| | **On-going projects C6** | | | |
| C06(92)1 459 | Risk assessment for sea and tidal defence structures. To define and indicate the use and understanding of probabilistic design of sea and tidal defence structures and to develop methods for assessing areas at risk for coastal and tidal flooding. | 2/93 3/96 | HR Wallingford | |
| C06(90)4 279 | Use of timber in sea defence schemes. To review available information on different types of timber and preservative. | 1/91 6/93 | Timber Research & Development Association | |
| C06(92)2 446 | Beach management manual. To produce a practical manual incorporating current practice and research findings to direct engineers on planning, design, implementation and management of beaches and beach recharge schemes. Supporting projects for research feeding into the manual will be: | 9/93 9/96 | CIRIA | |
| C06(92)13 441 | A. National database for monitoring data | 8/93 8/94 | Carl Bro Haiste | |
| C06(92)14 | B. Effectiveness of beach control operations | 2/93 9/93 | HR Wallingford | |
| C06(92)15 | C. Assessment of risks of beach recharge schemes | 94/95 94/95 | | |
| C06(92)16 489 | D. Use of non-aggregate materials | 9/93 9/95 | CIRIA | Contribution to CIRIA collaborative project |
| C06(92)3 480 | Saltmarsh management for Flood Defence. To produce practical guidelines and further advance the understanding of saltmarshes in sea defences and consequently enhance the NRA's capability in engineering and environmental management. Packages for research feeding into the guidelines will be: | 7/93 6/98 | Halcrow | |
| 444 | A. Saltmarsh management guide | 7/93 8/94 | Various | |
| | B. Maintenance and enhancement of saltings | 94/95 95/96 | | |
| | C. Historic changes in saltmarshes | 94/95 95/96 | | |
| | D. Experimental setback of saltings | 95/96 98/99 | | |
| | E. Estuary morphology | 94/95 95/96 | | |
| | F. Saltmarsh process | 94/95 97/98 | | |

| Proposal/ Project No. | Title | Start/ End | Contractor | Comments |
|---|---|---|---|---|
| C06(92)8 433 | Dissemination of Anglian Sea Defence Management Study<br>To produce practical guidance notes setting out the approach, methods and conclusions in generic terms which will benefit other regions. | 11/92 5/94 | In-house | |
| C06(92)7 522 | Public safety of access to coastal structures<br>To determine the accident record of different types of coastal structures, and to identify design features which enhance public safety while retaining the engineering effectiveness. | 2/94 9/95 | Halcrow | |
| | **Proposed new starts C6** | | | |
| C06(94)1 | Education and public awareness<br>To seek ways of explaining to the public why beach recharge and other soft defence methods are preferred. | 94/95 94/95 | | |
| C06(94)2 | Coastal management, including standards of performance for beaches<br>To identify the standard to which a beach should perform and how. | 94/95 96/97 | | |

## Outputs

This section lists the principal outputs that have been produced through the Topic Programme in 1993/94.

R&D Note 162 and R&D Project Records 242/5/A and 342/6/A	Stabilisation of soil embankments - Soil nailing

R&D Project Record 382/3/A	Armourstone foundations - Phase 1

R&D Note 197	Methodology for collating tidal water level data

R&D Project Record 386/4/A	Rehabilitation of coastal structures

R&D Project Record 366/3/A	Revetment systems and materials

## Summary information

The following summaries are provided for those documents available to the general public.

**NRA Publication Ref: R&D Note 162**

**Stabilisation of Soil Embankments - Soil Nailing**

1993

A F Howland Associates

NRA Project No. 342

Soil nailing is a means of improving the stability of soil slopes by rigid rods, or "nails", within a soil mass.

The established method of installing the nails is by drilling holes and grouting them in place. A variation of this has been developed by Soil Nailing Limited in conjunction with the University of Wales, College of Cardiff whereby the nails are placed by firing them into the ground from a compressed air gun. The main advantages of the fired soil nail lie in the speed and cost of installation. As the system requires less plant and support equipment than the drill and grout technique, the initial mobilization can be easier and provides a greater ability to accommodate any access restraints.

The fired soil nail has been used only on a limited number of occasions and is still undergoing some research and development both of the plant and equipment used and the design procedures necessary to support its use. The likelihood that the system of fired soil nailing becoming widely used in the near future in the United Kingdom is uncertain. Nonetheless, the National Rivers Authority should maintain its appraisal of the technique and an awareness of the potential in given situations.

Key words: Soil, embankment, flood defence.

**NRA Publication Ref: R&D Project Record 382/3/A**

**Armourstone Foundations - Phase 1**

Stickland I W (1993)

Posford Duvivier

NRA Project No. 382

The work investigated, by literature and method review, alternative types, methods and locations of constructions of foundations to rock armoured and rockfill structures, with particular reference to mobile foreshore conditions. The report gives an assessment of the requirements for an NRA engineers "guide" to the planning, design and construction of foundations to rock armour and rockfill structures.

The findings are summarised and conclusions and recommendations are given. The report concludes:

- the literature is insufficient to properly evaluate the nature and scale of the problems associated with armourstone foundations;

- obtaining the information will best be achieved by direct consultation with appropriate national authorities worldwide;

- there is concern that the required technology has not yet been fully established; and

- further work is needed to provide guidelines for practising engineers.

Key words: Breakwaters, rock groynes, reef breakwaters, foundations, failures and research.

**NRA Publication Ref: R&D Note 197**

**Methodology for Collating Water Level Data**

Parle P (1993)

Posford Duvivier

NRA Project No. 437

Guidance is given on the measurement, collection, storage, collation and analysis of tidal data. The greatest benefit of the development of a tidal database is likely to accrue to engineers involved in flood defence and flood forecasting. Further benefits will arise from increased certainty in decision-making in other areas such as development control, nature conservation and navigation.

The report also details the type of data that should be held on a tidal database, the Agencies which measure and store tidal data in the UK, a recommended method for storing the data, the tasks involved in the data transfer, the equipment which is needed, selection criteria for data retrieval and a discussion of the methods of tidal data analysis.

Key words: Tidal database, water level, software, hardware, format, data retrieval, analysis of tidal water level data.

### 5.5.6 Response to emergencies

**Topic Leader:**
Senaka Jayasinghe, Operations Manager (SE Area) with Thames Region.

**Rationale:**
The principal direction of the research in this Topic Area is towards the evaluation of the overall standards required of the NRA in flood emergencies. This work includes supporting the establishment of best methods of response, including flood and storm tide warnings and operation response procedures.

**Table 5.6 Response to Emergencies (Topic C8) R&D Programme (1994/95)**

| Proposal/ Project No. | Title | Start/ End | Contractor | Comments |
|---|---|---|---|---|
| | **On-going projects C8** | | | |
| C08(91)2 403 | Wave input to west coast storm tide model<br>To provide wave data to be used for the study of in-shore/ off-shore wave relationships for use in flood forecasting. | 5/92<br>6/93 | HR | |
| C08(91)3 431 | Emergency sealing of breaches<br>To investigate and develop alternative methods and materials for sealing breaches in NRA flood defences. | 4/94<br>10/94 | Posford Duvivier | |
| | **Proposed new starts C8** | | | |
| C08(94)1 | Benefit of flood warning and forecasting<br>To determine whether flood warning and forecasting systems are justified and the benefits of providing this service can be established. | 94/95<br>95/96 | | |

### Outputs

This section lists the principal outputs that have been produced through the Topic Programme in 1993/94.

R&D Project Record 289/2/T         Flood Defence emergency response
                                    - National levels of service

## 5.4 Commission C Budget

The R&D in Commission C is resourced through revenue from flood defence charges and the budget reflects the level of information, techniques and new methods required by the function to carry out its activities. Figure 5.1 shows the level of both actual and planned expenditure in this Commission. However, the NRA is not the only organisation with a need to carry out R&D in this area. Part-funded work with the Institute of Hydrology, Ministry of Agriculture, Fisheries and Food, Crown Estates and others ensure that the NRA maximises the use of its resources in this area.

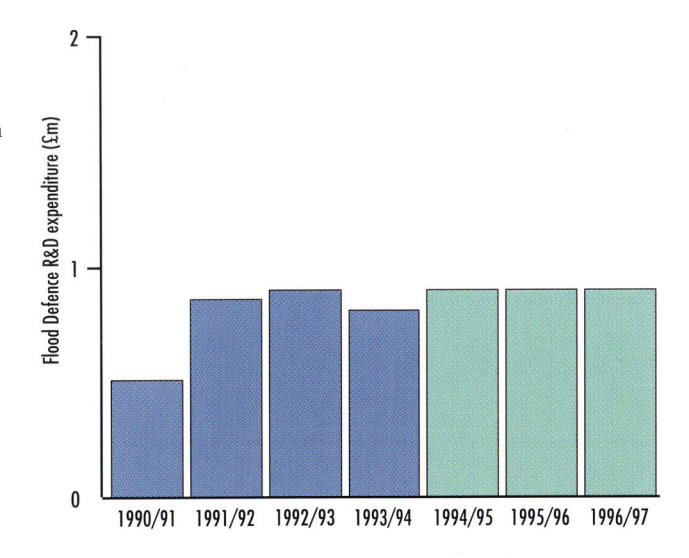

**Figure 5.1** Actual and planned expenditure on Flood Defence R&D up to the financial year 1996/97

# 6. FISHERIES

This section sets out the progress made in the Commission D R&D Programme in 1993/94. Also described here are the Topic Programmes for the 1994/95 financial year, together with the strategic direction of the R&D required in support of the Fisheries function. This area of R&D is overseen by Chris Mills - Area Fisheries, Conservation and Recreation Manager in North West Region - as Commissioner.

## 6.1 Business Rationale

The NRA published its Fisheries Strategy in 1993, setting out its objectives and key activities over a ten-year horizon. The three key areas where R&D is required in order to support function activities are:

- Resource assessment;
- Environmental and biological pressures; and
- Fisheries management and stock assessment.

The NRA has a stated aim to monitor the fisheries status of surface waters. An important component of this is the assessment of resources, both at an individual species level as well as at a population level. This information on the biology of key species such as eels, sea trout and salmon, is essential for the operational management of fisheries, and is provided through a business-focused R&D Programme.

In order to be able to interpret monitoring data and to carry out effective fisheries management in response to either natural or man-made changes, a considerable understanding of environmental and biological influences on fish populations is essential. Such influences can range from disease to changes in land use, as well as the effects of other fish species.

The development and implementation of effective stock assessment and management procedures will continue to be required if the NRA is to be successful in achieving its aim of maintaining, improving and developing fisheries in England and Wales. The development of new stock assessment techniques, in particular, will continue to feature strongly in the R&D Programme.

## 6.2 Outputs Produced

In support of the NRA's resources assessment activities, R&D projects provided a series of outputs during 1993/94. These included information on eel and elver stocks in the Severn Estuary and an assessment of the rare fish in lakes in England and Wales. A classification scheme for fisheries was also produced and is in the process of being evaluated by NRA staff. This scheme is designed to provide a clear analysis of stock assessment data and to provide an assessment of the performance of operational activities.

An assessment of the body burdens of chemicals in fish was also reported during 1993/94. Fish are often useful indicators of water quality, as this affects the health of populations.

A draft procedure for the design and evaluation of fish passes has been produced and is being trialled by NRA fisheries staff. The procedures will be reassessed following this trial period, and will subsequently be revised before being put on a wider distribution.

## 6.3 Scientific Rationale

The Fisheries function within the NRA is highly dependent upon solid and up-to-date scientific information in order to carry out its core activities. The review of the status of rare fish species in England and Wales is a prime example of the NRA's interest in understanding the current pressures on resources.

Further areas where the Commission D programme is helping to introduce new scientific and technological approaches include the development of hydroacoustic methods of fish survey; new information on the ecology and genetic variation of rare fish; and an insight into the migratory behaviour of salmon and sea trout smolts in estuaries.

## 6.4 Strategic Projects

There is a range of projects within Commission D which underpin aspects of the NRA Fisheries Strategy. In 1993/94, a project looking at the current information and stock assessment capabilities regarding sea trout was commenced. This work will define the prime issues in this area, following which an effective programme of investigations will be set out in order to enable the NRA to manage sea trout stocks in a sustainable manner.

An important collaborative project with MAFF has been established to develop fish tracking devices. In particular, this work will look at tags and data logging systems, with the aim of providing further information on fish biology.

The project on coarse fish populations in lowland rivers has reached the end of the initial phase to identify issues in which further research could assist course fish management. The NRA is currently reviewing which elements to pursue in further phases of this strategically important area.

## 6.5 Topic Programmes and Outputs

This section covers the rationale behind the projects undertaken in each of the three Topic Areas within the Fisheries Commission. The outputs that have been produced between April 1993 and May 1994 are also summarised.

Summary information is provided for those outputs available to the general public; a complete listing of all outputs together with their availability is provided as an insert in the back cover.

Where an output is still under consideration by the NRA, where it contains commercial in confidence information, or where it feeds into another activity such as the development of a functional manual, the current availability may be less widespread. External organisations wishing to know more about such outputs should contact the relevant Topic Leader.

The R&D Programme for each Topic Area is provided in Tables 6.1 to 6.3. These tables list the projects either on-going or proposed for starting in the financial year 1993/94. Each Regional R&D Coordinator and the R&D Section at Head Office has further details on all such projects.

A fold-out guide to the information contained in the following sections is provided on the inside front cover.

### 6.5.1 Fisheries Resources

**Topic Leader:**
Alan Winstone, Senior Fisheries Scientist with Welsh Region.

**Rationale:**
The overall objective for work within the Topic Area is to provide information on the biology of individual fish or fish populations in order to derive basic resource information for operational management.

**Table 6.1 Fisheries Resources (Topic D1) R&D Programme (1994/95)**

| Proposal/ Project No. | Title | Start/ End | Contractor | Comments |
|---|---|---|---|---|
| | **On-going projects D1** | | | |
| D01(90)1 256 | Eel and elver stock assessment<br>To assess elver stocks in the Severn Estuary together with elver exploitation and its implications for river stocks. | 11/90 10/93 | Polytechnic of Central London | |
| D01(90)3 244 | Development of a fisheries classification scheme<br>To enable clear analysis of stock assessment data used for operational duties and to assess performance. | 10/90 3/94 | WRc | |
| D01(91)2 404 | Use of catch statistics to determine fish stock size<br>To evaluate the use of migratory salmonid, eel and freshwater fish catch statistics for the management of these stocks in England and Wales and to determine how they can best be used to estimate stock size. | 4/92 4/95 | EAU | |
| D01(91)3 443 | Sea trout investigation<br>To review and evaluate current knowledge, research and stock assessment capability in relation to sea trout and to design a cost effective programme of investigations which will enable the NRA to effectively manage sea trout stocks sustainably. | 1/93 5/94 94/95 95/96 | D Solomon | |
| D01(92)2 438 | Genetic aspects of spring run salmon<br>To review information on the genetic characteristics and time of return of spring salmon and to make recommendations on the feasibility of, and techniques for, enhancement of spring salmon. | 1/93 5/93 | University College, Cork | |
| D03(93)1 500 | Effectiveness of salmonid stocking strategy<br>To identify the most cost-effective strategies for stocking migratory salmonids in order to maximise returns of adult fish to fisheries and/or their natal river. | 11/93 5/94 | Fishskill | |

| Proposal/ Project No. | Title | Start/ End | Contractor | Comments |
|---|---|---|---|---|
| | **Proposed new starts D1** | | | |
| D01(94)1 | Classification of river fisheries<br>To classify, by computer program, all waters on the Fisheries Database and to produce a computer program for the use of all Fisheries Staff in future classifications. | 94/95<br>94/95 | | |

## Outputs

No outputs were produced through the Topic Programme in 1993/94.

### 6.5.2 Environmental and Biological Influences

**Topic Leader:**
Nigel Milner, Area Environmental Appraisal Manager with Welsh Region.

**Rationale:**
The provision of improved understanding of the effects of environmental and biological factors on fish populations is the basis for R&D in this Topic Area. This will enable monitoring programmes to be correctly interpreted and effective management procedures to be introduced where changes occur.

**Table 6.2 Environmental and Biological Influences (Topic D2) R&D Programme (1994/95)**

| Proposal/ Project No. | Title | Start/ End | Contractor | Comments |
|---|---|---|---|---|
| | **On-going projects D2** | | | |
| 152 | River quality and fisheries status (Torridge)<br>To quantify the effects of land use change on river quality and fisheries through a case study on the River Torridge catchment. | 1/90<br>10/93 | WRc | Part-funded by MAFF |
| D02(90)2<br>229 | Disease status of fish as an additional indicator of water quality<br>To evaluate the disease status of fish as an indicator of surface water quality, and the role of fish health factors in limiting population success. | 10/90<br>10/93 | In-house | |
| D02(90)7<br>312 | Estuarine migratory behaviour of salmon and sea trout smolts<br>To investigate, using advanced telemetry techniques, and describe the behaviour of Atlantic salmon and sea trout smolts during the estuarine phase of their migration. | 1/94<br>10/95 | In-house | |
| D02(91)2<br>338 | Habscore<br>To develop a stream habitat procedure that will enable prediction of salmon abundance from stream features; and which will assist the NRA to achieve related statutory objectives. | 7/91<br>3/94 | WRc | |
| D02(91)1<br>429 | Coarse fish populations in large lowland rivers<br>To identify the critical factors constraining coarse fish populations in lowland rivers and to determine management strategies to develop and improve the fisheries. | 11/92<br>3/94 | Humberside International Fisheries Institute | |
| D02(92)2<br>452 | Effects of stocked trout on the survival of wild fish populations<br>To assess the effect of stocked trout on the survival of wild fish stocks in order to produce recommendations on future trout stocking policies. | 1/92<br>9/95 | WRc | |
| | **Proposed new starts D2** | | | |
| D02(94)1 | Setting of spawn biomass targets<br>To establish the optimum spawn biomass to maintain self-sustaining salmon and sea trout stocks to underpin the national migratory fisheries management plan. | 94/95<br>94/95 | | |

## Outputs

This section lists the principal outputs that have been produced through the Topic Programme in 1993/94.

R&D Note 193 and                                Body burdens in fish
R&D Project Record 397/10/ST

R&D Note 40                                     Strategic ecosystem studies of slow-flowing lowland rivers

## Summary information

The following summaries are provided for those documents available to the general public.

**NRA Publication Ref: R&D Note 40**

**Strategic Ecosystem Studies of Slow-Flowing Lowland Rivers**

Pinder L C V *et al.* (1992)

Institute of Freshwater Ecology

NRA Project No. 123

The Great Ouse river was chosen for these studies largely because of a reported decline in the quality of the fishery over recent decades. This decline has affected bream populations particularly strongly. Preliminary sampling during the first year of the study indicated that physical conditions, rather than water chemistry, were most likely to be having a deleterious effect on the fishery at the present time. It is obvious that unless suitable physical conditions for fish and other aquatic organisms exist, the ecological benefits of improved water quality will not be fully realised.

In the main river channels of the Great Ouse the development of plants is restricted by a combination of channel morphology and water turbidity as well as by dredging and periodic cutting and removal. At high discharge levels very little of the river is suitable for small cyprinids which are able to maintain station only at velocities of a few centimetres per second. In these circumstances marinas and back-channels with low current velocity should be given to the connection of many more disused gravel pits to the river system. These would not only increase the extent of the available refuges for fish but, with appropriate management, would also provide more extensive spawning sites for species like bream, which have declined particularly noticeably and have a restricted distribution in the river at present.

Key words:     Ecosystems, fish, algae, water quality, river management.

### 6.5.3 Fisheries Management

**Topic Leader:**
Steve Bailey, Fisheries Recreation, Conservation and Navigation Co-ordinator with Northumbria & Yorkshire Region.

**Rationale:**
The maintenance, improvement and development of fisheries requires effective management strategies. This Topic Area provides information for the development of such strategies that are cost-effective.

**Table 6.3 Fisheries Management (Topic D3) R&D Programme (1994/95)**

| Proposal/ Project No. | Title | Start/ End | Contractor | Comments |
|---|---|---|---|---|
| | **On-going projects D3** | | | |
| D01(90)2 249 | Status of rare fish<br>To determine the present status of rare fish in certain lakes of England and Wales, and to compile related information on the ecology and genetic variation of these species which is necessary to safeguard their populations. | 10/90<br>9/94 | IFE | |
| D03(90)2 250 | Surveying by hydroacoustic techniques<br>To evaluate and develop the use of hydroacoustic techniques to quantitatively estimate the abundance and size distribution of fish in larger watercourses.<br>Phase 2 - production and use of field manual | 5/94<br>9/94 | In-house/Royal Holloway and Bedford New College | |
| D03(91)1 325 | Development of fish stock assessment methodologies and methods<br>To evaluate and develop efficient fishery survey strategies and methodologies and produce practical guidelines for their implementation in NRA stock assessments. | 7/91<br>5/94 | WRc | |
| D03(90)1 370 | Design and use of fish counters and radio-telemetry<br>To evaluate the performance of automatic fish counters and their use in monitoring adult migratory salmonid stocks. | 7/91<br>9/94 | In-house | |
| D03(91)2 334 | Electric fishing of deep rivers<br>To develop electric fishing sampling equipment and methodology which will allow appraisal of fish stocks in large lowland rivers. | 10/91<br>3/94 | Humberside International Fisheries Institute | |
| D03(92)3 440 | Survival and dispersal of stocked coarse fish<br>To evaluate the effectiveness of stocking with coarse fish, both hatchery-reared and transferred from other waters, and to determine the most effective stocking practices. | 10/92<br>4/93<br>94/95<br>96/97 | WRc | |
| D03(93)3 486 | Assessing salmon stocks using a hydroacoustic counter<br>To install, operate and evaluate a hydroacoustic fish counter on the River Wye in order to produce reliable data for stock management. | 2/94<br>3/97 | In-house | Part OI/part R&D links to Project 250 |
| D03(92)1 503 | Fish tracking developments<br>To undertake a definition study to appraise the options to develop fish tracking equipment, in particular tags and data logging systems, in order to improve the efficiency of NRA tracking studies and to obtain a greater understanding of fish biology. | 12/94<br>4/94<br>94/95<br>95/96 | MAFF | Collaborative project with MAFF |
| D03(93)4 511 | Effects of retention of fish in keepnets<br>To establish the extent of physiological perturbations in fish arising from capture and confinement in a keepnet. | 1/94<br>3/94 | IFE | Part-funded by National Federation of Anglers (NFA) |
| | **Proposed new starts D3** | | | |
| D03(92)2 | Methods for fishery rehabilitation<br>To develop new methods of environmental restoration (other than water quality related) that will enable the NRA to improve, develop and rehabilitate damaged fisheries. | 94/95<br>96/97 | | |

| Proposal/<br>Project No. | Title | Start/<br>End | Contractor | Comments |
|---|---|---|---|---|
| D03(93)2 | Development of systems for designing and implementing surveys<br>To develop systems for designing and implementing fisheries survey, including evaluation of alternative strategies. | 94/95<br>95/96 | | |
| D03(94)1 | Smolt trapping using acoustic deflectors<br>To investigate how sound can be best used to direct the route for migration in order to increase trapping efficiency. | 94/95<br>95/96 | | |

## Outputs

This section lists the principal output that has been produced through the Topic Programme in 1993/94.

R&D Report 5    Fish pass design and evaluation (to be revised as an R&D Note in 1994/95)

## 6.6  Commission D Budget

The R&D in Commission D is resourced through grant-in-aid and the budget reflects the level of information, techniques and new methods required by the function to carry out its activities as well as the number of projects the function feels able to support. Figure 6.1 shows the level of both actual and planned expenditure in this Commission. However, the NRA is not the only organisation with a need to carry out R&D in these areas. Part-funded work with the Institute of Freshwater Ecology and the Ministry of Agriculture, Fisheries and Food is already underway.

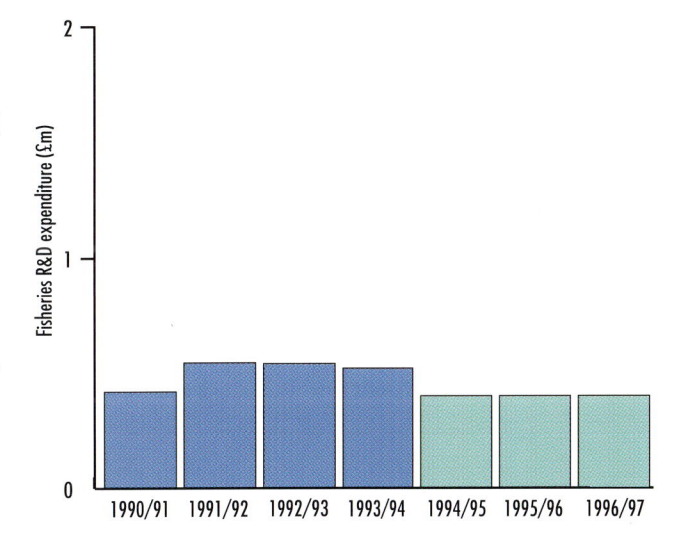

**Figure 6.1 Actual and planned expenditure on Fisheries R&D up to the financial year 1996/97**

# 7. RECREATION AND NAVIGATION

This section sets out the progress made in the Commission E R&D Programme in 1993/94. Also described here, are the Topic Programmes for the 1994/95 financial year, together with the strategic direction of the R&D required in support of the Recreation and Navigation functions. This area of R&D is overseen by Paul Raven, Conservation Officer at Head Office, as Commissioner.

## 7.1 Business Rationale

The NRA published strategies for both Recreation and Navigation in 1993, setting out their objectives and key activities over a ten year horizon. The key areas where R&D is required in order to support function activities are:

**Recreation**

- Assessing and monitoring;
- Operations; and
- Collaboration.

**Navigation**

- Regulation of NRA navigations;
- Operations; and
- Improvement of navigations in England and Wales.

The NRA aims to develop and manage its own sites and provide facilities such as canoe slaloms, boat moorings and riverside footpaths for the benefit of both the environment and the general public where appropriate. R&D will provide a key role in supplying supporting knowledge in this area.

An important consideration in the effective management of the recreation potential of surface waters is the availability of water of sufficient quality and quantity for the safe promotion of recreation activities. The NRA also ensures that the effects of these activities on the environment itself are given due consideration. The information necessary in balancing such requirements is provided by the R&D Programme, and in certain areas through collaboration with the programmes of other bodies such as British Waterways and the Broads Authority.

As part of the NRA's management of its navigations and sites, information is required on trends and demands in participation, as well as regular monitoring of actual levels of usage. In addition, the NRA aims to promote the use of new boat technology, where this may have beneficial environmental effects.

## 7.2 Outputs Produced

The number of outputs produced through the Commission E programme reflects the overall level of work in this area. In 1993/94 no completed outputs were delivered, but a number of important draft reports was presented to the customer for consideration.

An assessment of the field trials of various approaches to bank protection was completed in 1993, with the output being held over until a comprehensive manual can be produced in collaboration with British Waterways and the Broads Authority.

Initial information from an NOP survey of the socio-economic aspects of angling has been produced, and the NRA is aiming to publish this work in 1994 as a Recreation Report.

## 7.3 Scientific Rationale

The main thrust of the R&D Programme in this area is to improve the efficiency and effectiveness of the NRA, and to underpin policy development.

With this in mind, there are two principal approaches to introducing new science into the NRA. Firstly, by assessing participation trends through targeted surveys and public perception assessments, the NRA hopes to ensure that decisions are based on sound social and economic principles. In addition, where practical improvements are required to river channels in order to enhance either the recreation or navigation potential, the identification, evaluation and subsequent introduction of best current practice is essential. In relation to bank erosion, this has involved assessing the aesthetic appeal of remedial techniques, as well as fundamental aspects of bank stability.

## 7.4 Strategic Projects

Commission E is comparatively small but relies on well targeted projects which can provide outputs for use in day-to-day operations.

The socio-economic review of angling, started in late 1993, will provide information to underpin decision-making in fisheries and recreation.

Work is still continuing on the impact of recreation on wildlife. The NRA has duties for both recreation and conservation, and this will assist policy-making and site management where a balance of these interests is required. The initial work will be completed in 1994, although the strategic nature of this issue will dictate the need for further R&D in this area.

## 7.5 Topic Programme and Outputs

This section covers the rationale behind the projects undertaken in the Recreation and Navigation Commission. The outputs that have been produced between April 1993 and May 1994 are also included.

Summary information is provided for those outputs available to the general public; a complete listing of all outputs together with their availability is provided as an insert in the back cover.

Where an output is still under consideration by the NRA, where it contains commercial in confidence information, or where it feeds into another activity such as the development of a functional manual, the current availability may be less widespread. External organisations wishing to know more about such outputs should contact the relevant Topic Leader.

The R&D Programme for the sole Topic Area in this Commission is provided in Table 7.1. This table lists the projects either on-going or proposed for starting in the financial year 1994/95. Each Regional R&D Coordinator and the R&D Section at Head Office has further details on all such projects.

A fold-out guide to the information contained in the following sections is provided on the inside front cover.

### 7.5.1 Recreation and Navigation

**Topic Leader:**
Craig McGarvey, Recreation and Navigation Officer with Head Office.

**Rationale:**
The prime area of work for this Topic Area is to develop specific guidelines to enable the NRA to promote the navigation amenity and recreational potential of surface waters and associated lands. The work regarding navigation is restricted to waters where the NRA is the controlling authority.

**Table 7.1 Recreation and Navigation (Topic E1) R&D Programme (1994/95)**

| Proposal/ Project No. | Title | Start/ End | Contractor | Comments |
|---|---|---|---|---|
| | **On-going projects E1** | | | |
| E01(91)2 336 | Bank erosion on navigable waterways  To conduct field trials to identify the most appropriate technique of bank erosion protection for navigable rivers under NRA responsibility. | 7/91 7/93 | University of Nottingham | Follow on from Project 225 |
| E01(91)1 498 | Impact of recreation on wildlife  To provide information to enable the NRA to reconcile its recreation and conservation duties - initial review.  Phase 2 - more detailed studies | 11/93 6/94  94/95 95/96 | Land Use Consultants | Liaison with SNH |
| E01(93)1 501 | Socio-economic review of angling  To assess the current status of angling in England and Wales and analyse the distribution of NRA rod licence holders to enable the NRA to plan and forecast income and research implications. | 11/93 4/94 | NOP | |
| | **Proposed new starts E1** | | | |
| E01(94)2 | Demand studies - boating on NRA navigations  To develop an understanding of boating demand to assist the setting of charges, price structures and prioritisation of capital and revenue works. | 94/95 94/95 | | Collaboration with BWB |
| E01(94)3 | Use value of NRA navigations  To apply "contingent valuation" methods to NRA navigations to identify visitor value to enable the NRA to better quantify benefits of the navigation function. | 94/95 94/95 | | Collaboration with BWB |

**Outputs**

No outputs were produced through this Topic programme in 1993/94.

## 7.6  Commission E Budget

The budget for Commission E projects is raised from grant-in-aid. It is relatively low-level, when compared to other Commissions, and is a reflection of both the number of projects the function can manage together with the need for R&D to support the functional initiatives and continuing activities. Figure 7.1 illustrates the planned and actual expenditure on recreation and navigation R&D up to 1996/97.

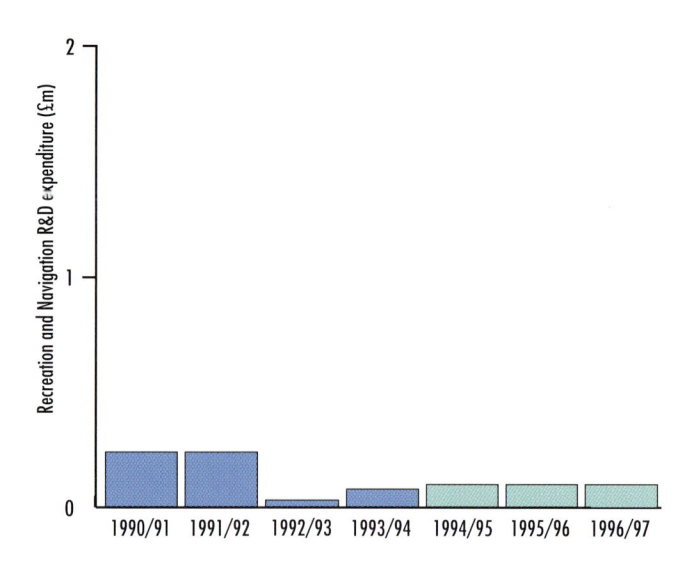

**Figure 7.1** Actual and planned expenditure on Recreation and Navigation R&D up to the financial year 1996/97

# 8. CONSERVATION

This section sets out the progress made in the Commission F R&D Programme in 1993/94. Also described here are the Topic Programmes for the 1994/95 financial year, together with the strategic direction of the R&D required in support of the Conservation function. This area of R&D is overseen by Paul Raven, Conservation Officer at Head Office, as Commissioner.

## 8.1 Business Rationale

The NRA published its Conservation Strategy in 1993, setting out its objectives and key activities over a ten year horizon. The NRA has a stated aim in relation to conservation:

> "to conserve and enhance wildlife, landscape and archaeological features associated with inland and coastal waters of England and Wales"

This will be achieved either directly, through the Authority's own operational and regulatory affairs, or by influencing the activities of others.

Responsible management of the natural environment relies on effectively assessing its value, identifying those factors which may impact on this resource and ultimately the management strategies that reflect the need to target the NRA's operations in a sustainable manner. In order to support this overall approach, the business rationale for this function is underpinned through R&D in two key areas, namely:

- Conservation resource appraisal and impact assessment; and
- Conservation management.

As many of the improvements to the conservation value of the water environment can be made through the activities of other NRA functions, close liaison is necessary with the R&D Programmes being pursued through other Commissions.

## 8.2 Outputs Produced

The issue of the predation of fish by birds such as cormorants, goosanders and red-breasted mergansers has been an important one for the NRA. As part of support to policy development in this area, a number of outputs were produced during 1993/94, with R&D Report 15 entitled "Fish Eating Birds" being published in early 1994. This report set out the results of the R&D project as well as the NRA's position statement on cormorants and sawbill ducks.

An assessment of the NRA's legal responsibilities in relation to conservation in coastal areas was reported in another output delivered during 1993. This work compared the NRA's responsibilities with those of other organisations involved in conservation activities, and will be developed further as part of a concerted programme of conserving coastal habitats.

A key link to the work of Flood Defence engineers and Water Quality scientists was the subject of another output. This work assessed the physical environment necessary to support invertebrate communities and developed a unified method for the ecological assessment of "functional habitat" which can be used by other NRA staff.

## 8.3 Scientific Rationale

Conservation-related activities require a fundamental understanding of the factors affecting various aspects of the water environment. This understanding needs to be based on sound scientific principles.

The protection of freshwater crayfish, for instance, has relied heavily on an appreciation of the effects of introducing alien species into rivers as well as the impact of crayfish plague on natural populations.

The factors affecting wetland areas in England and Wales are being investigated in order to achieve an effective wetland management strategy. This involves understanding the role of water levels, particularly with regard to sustainable creation and management of wetland habitats.

## 8.4 Strategic Projects

The NRA has a duty with respect to the impact of its activities on archaeological sites. An initial phase of a strategic project designed to review the NRA's approach towards such sites and discoveries has been developed. The outputs from this project will help the NRA in assessing the potential environmental impact of its own activities.

Two related projects have been developed to provide a classification procedure for river habitat survey data. The first of these will look at calibrating and validating the River Habitat Survey (RHS) methodology through independent sampling of river and stream sites as well as quality checking NRA collected data. The second project will analyse RHS data with a view to producing a national river typology and habitat quality classification. The procedures developed through these projects will be implemented through a national methodology in 1995/96.

## 8.5 Topic Programmes and Outputs

This section covers the rationale behind the projects undertaken in each of the three Topic Areas within the Conservation Commission. The outputs that have been produced between April 1993 and May 1994 are also summarised.

Summary information is provided for those outputs available to the general public; a complete listing of all outputs together with their availability is provided as an insert in the back cover.

Where an output is still under consideration by the NRA, where it contains commercial in confidence information, or where it feeds into another activity such as the development of a functional manual, its current availability may be less widespread. External organisations wishing to know more about such outputs should contact the relevant Topic Leader.

The R&D Programme for each Topic is provided in Tables 8.1 and 8.2. These tables list the projects either on-going or proposed for starting in the financial year 1994/95. Each Regional R&D Coordinator and the R&D Section at Head Office have further details on all such projects.

A fold-out guide to the information contained in the following sections is provided on the inside front cover.

### 8.5.1 Conservation Resource Appraisal and Impact Assessment

**Topic Leader:**
Peter Barham, Regional Conservation Officer, Fisheries and Recreation Manager with Anglian Region.

**Rationale:**
The principal aim of this Topic Programme is to develop effective methods for surveying, monitoring and classifying the conservation resource. The R&D will include the evaluation of impacts on the resource as well as defining acceptable criteria for monitoring and enhancing the conservation status.

**Table 8.1 Conservation Resource Appraisal and Impact Assessment (Topic F1) R&D Programme (1994/95)**

| Proposal/ Project No. | Title | Start/ End | Contractor | Comments |
|---|---|---|---|---|
| | **On-going projects F1** | | | |
| F01(90)25 294 | Control of invasive riparian and aquatic weeds. To control the spread of invasive plants and to formulate and promote good management practices which would enhance the conservation and amenity value of areas for which the NRA is responsible. | 2/91 5/94 | ICOLE | |
| F01(92)2 418 | Impact assessment and acceptable conservation criteria. To evaluate impacts upon and identify acceptable criteria for maintaining and enhancing the conservation status of inland and coastal waters and associated land in England and Wales. | 94/95 95/96 | | |
| F01(92)4 477 | Rehabilitation of degraded habitats. To determine the effects of land use and river management practices on riverine habitats so as to more objectively define and implement an effective rehabilitation programme for degraded river corridors. | 6/93 3/94 94/95 95/96 | ICOLE | Links to Project 340 |
| F01(93)1 506 | Impact of NRA activities on archaeology. To review the duties of the NRA to archaeology and the approach to be taken towards archaeological sites and discoveries, in order to enable effective execution of the NRA's statutory responsibilities in this area. | 1/94 3/94 94/95 95/96 | In-house | |
| F01(93)3 517 | River morphology and plant relationships. To establish key relationships between physical characteristics of river types and associated aquatic plant communities by analysing data collected by NCC. | 2/94 5/94 | T Rowell | |

| Proposal/ Project No. | Title | Start/ End | Contractor | Comments |
|---|---|---|---|---|
| F01(94)2 526 | Validation and calibration of River Habitat Survey data<br>To calibrate and validate River Habitat Survey (RHS) methodology by sampling river and stream sites in England and Wales and quality-checking selected sites being surveyed by NRA staff. | 5/94<br>9/94 | IFE | Links to Project 469 |
| F01(94)5 530 | Working classification of River Habitat Survey<br>To analyse RHS field survey data and generate a national river typology and habitat quality classification in order to produce a national methodology to be implemented in 1995/96. | 6/94<br>10/94 | IFE | |
| **Proposed new starts F1** | | | | |
| F01(93)2 | Conservation assessment and management strategies for ponds and other small water bodies<br>To develop a cost-effective, predictive means of assessing the conservation status of ponds and to identify management techniques which optimise aquatic and associated terrestrial habitats. | 94/95<br>95/96 | | |
| F01(94)1 | River channel typology for catchment and river management<br>To develop a standard method of surveying channel cross-sections with respect to form an associated vegetation/habitat structure, in a fashion amenable to easy data management and computerisation. | 94/95<br>94/95 | | Link with Topic C1 and C01(91)3 |
| F01(94)4 | Investigation of the distribution and impact of a fungal disease of alders<br>To establish current and future distribution of the disease, identify potential impacts on the river corridor and recommend management options for alleviating impacts. | 94/95<br>94/95 | | |

## Outputs

This section lists the principal outputs that have been produced through the Topic Programme in 1993/94.

R&D Note 107                              Review of nature conservation survey methodologies

## Summary information

The following summaries are provided for those documents available to the general public.

**NRA Publication Ref: R&D Note 107**

**Review of Nature Conservation Survey Methodologies**

1993

Environmental Advisory Unit

NRA Project No. 393

This report summarises the ecological survey methods which are used to assess the nature conservation value of sites. The aim was to recommend survey methods for use by the NRA to fulfil its responsibilities in nature conservation. In addition, a computer database of organisations holding ecological survey information was compiled. The study covered both freshwater and coastal habitats within the NRA's remit.

The Phase 1 (extensive, habitat-level), followed by phase 2 (intensive, species-level) approach to ecological surveys is accepted as best practice. It is advised that intermediate types of survey method, equivalent to river corridor surveys, are not suitable for consistent conservation evaluation.

The report summarises standard or recommended Phase 1 and 2 methods for flora and fauna. These should be used for nature conservation surveys by the NRA. The database should be consulted for sources of existing survey data and information on distribution of habitats and species.

Key words: Conservation, ecological survey, biological data, survey methods, habitats, species, database.

### 8.5.2 Conservation Management

**Topic Leader:**
Richard Howell, Conservation and Recreation Officer with Welsh Region.

**Rationale:**
The development of management strategies and techniques for protecting the conservation resource, will form the basis of R&D in this Topic Area. They will be designed such that they are compatible with the other duties of the NRA.

**Table 8.2 Conservation Management (Topic F2) R&D Programme (1994/95)**

| Proposal/ Project No. | Title | Start/ End | Contractor | Comments |
|---|---|---|---|---|
| | **On-going projects F2** | | | |
| F01(91)11 400 | Appraisal of conservation enhancement of flood defence works  To develop a method for post-project appraisal of habitat conservation and enhancement works and to assess the value of such works in relation to natural recovery from NRA operational schemes. | 3/92 3/95 | University of Loughborough | Follows from Project 285 |
| F01(91)10 352 | Aquatic flora database  To develop and update a computerised database of submerged and floating aquatic vascular plants in order to provide up to date information on their occurence and distribution. | 10/91 3/95 | ITE | |
| F01(91)1 378 | Conservation of freshwater crayfish  To assess the impact of introductions of non-native crayfish and outbreaks of crayfish plague on freshwater ecosystems and to formulate a strategy for the conservation of the native species (*Austropotamobius pallipes*). | 1/92 12/94 | University of Nottingham | Technical collaboration with English Nature and MAFF |
| F01(91)2 383 | Wetland creation/river corridor enhancement  To assess the conservation value of 'off-river' wetland habitats created during a river corridor enhancement programme and investigate experimental management techniques for increasing the native conservation value of created habitats. | 1/92 1/95 | Pond Action | Part OI/part R&D. Links with Topics B2, C1 and C3 |
| A08(91)5 346 | Physical environment for river invertebrate communities  To develop a unified method for the ecological assessment of "functional habitat" analysis for use by Flood Defence engineers and Water Quality and Conservation scientists. | 4/91 3/94 | University of Leicester | Phase 2 of Part OI/Part R&D |
| F02(92)3 461 | Species management in aquatic habitats  To identify effective management strategies for protecting and enhancing the status of rare or dispersed species and minimising the impact of nuisance species, associated with inland and coastal waters in England and Wales. | 3/93 9/93 94/95 95/96 | WRc/ITE | |
| A02(91)1 313 | Classification of catchment sources in relation to the ecological effects of wetland liming  To refine the understanding of the potential biological effects of catchment liming on important conservation resources and produce a classification system for upland wetlands of different conservation value and susceptibility. | 7/91 5/94 | University of Wales, Cardiff | Part-funded by CCW |
| F02(92)1 472 | Conservation in coastal areas  To clearly define and interpret NRA legal responsibilities in the coastal zone and relate these to the legal responsibilities of other relevant organisations and to NRA's own conservation duties. | 5/93 8/93 94/95 95/96 | University of Wales, Cardiff | |
| F02(92)2 474 | Wetland conservation  To identify the wetland resource and develop effective management strategies to define the NRA's role in the protection, rehabilitation and creation of wetlands in England and Wales. | 5/93 3/94 94/95 95/96 | ECUS | |

| Proposal/ Project No. | Title | Start/ End | Contractor | Comments |
|---|---|---|---|---|
| F02(93)2 525 | **Survey of habitats for invertebrates on exposed riverside sediments** To review information obtained from river corridor habitat and other NRA surveys on habitat features in order to assess their likely importance. | 5/94 2/95 | Entomological Monitoring Services | |
| | **No proposed new starts F2** | | | |

## Outputs

This section lists the principal outputs that have been produced through the Topic Programme in 1993/94.

R&D Report 15    Fish-eating birds
 - Assessing their impact on freshwater fisheries

R&D Note 267 and    Avian Piscivores: Basis for policy
R&D Project Record
461/8/NA&Y

## Summary Information

The following summaries are provided for those documents available to the general public.

**NRA Publication Ref: R&D Report 15**

**Fish-Eating Birds - Assessing their Impact on Freshwater Fisheries**

Marquiss M and Carss D N (1994)

Institute of Terrestrial Ecology

NRA Project No. 461

This work reviewed the existing information on the status of fish-eating birds, their impacts on fisheries and the effectiveness of management procedures to control bird predation on fish populations. The study also discusses criteria for serious damage to fisheries, evaluates the current NRA position statement and suggests future action with regard to policy and research.

Cormorants have increased dramatically in Europe and to some extent in England and Wales. Their increasing use of inland freshwaters have coincided with increased fish farming and intensive fish stocking for angling. Newly established breeding populations have increased, particularly in southeast England. There are good reasons to anticipate cormorant numbers eventually stabilising, but the process will be protracted as long as their food supply continues to be artificially increased. Sawbill duck numbers have also increased in parallel with an expansion of their breeding range. However, this expansion has now slowed and numbers have fluctuated, perhaps even declined, within the existing range.

Studies of diet suggest that all three species are unlikely to be damaging fisheries over much of their range. Experimental fish community studies show that predation can affect fish populations, but to date there are no experimental studies that have shown bird predation to damage fisheries seriously other than at small enclosed systems such as fish farms. Most evidence of damage is anecdotal or circumstantial. Models of bird/fish interaction have so far been unsatisfactory and their predictions remain untested. The last of hard evidence for bird damage to fisheries is not necessarily because there is no bird effect, but could be because the appropriate experiments have not yet been done. No study has shown that killing birds effectively prevents damage to fisheries and moreover, work currently underlay in Britain is unlikely to do so.

Key words: Cormorant, sawbill duck, fish-eating birds, fisheries, impact.

**NRA Publication Ref: R&D Note 267**

**Avian Piscivores: Basis for Policy**

Marquiss M and Carss D N (1994)

Institute of Terrestrial Ecology

NRA Project No. 461

In Britain and elsewhere, fish-eating birds, particularly cormorants and sawbill ducks (goosanders and red-breasted mergansers), are widely believed to affect fisheries, reducing the abundance of fish or changing their behaviour as to reduce harvest or angling catches. These birds are protected by law but with provision for the issue of licences (by SOAFD in Scotland, MAFF in England and WOAD in Wales) to kill them to "prevent serious damage to.. fisheries". There is considerable debate as to whether these birds damage fisheries, and whether licensed killing effectively prevents serious damage.

The review concludes with recommendations that:

1. the current NRA position, of not supporting licensed killing until serious damage has been established and killing proven to be the most effective management procedure for preventing it, remains unchanged until new information accrues;

2. the NRA initiates in a supporting role, research to address the problem using an experimental approach;

3. experimental research should investigate whether bird predation has a measurable impact on fisheries, and if so, what mechanisms are the most cost-effective in reducing this impact; and

4. fisheries interests, the licensing authorities, nature conservation organisations and the NRA a have common interest in promoting the appropriate research and should consider joint funding.

Key words: Fish, predation, birds, management.

NRA Publication ref: R&D Project Record 461/08/N&Y provides further details in support of R&D Note 267.

## 8.6 Commission F Budget

The R&D in Commission F is resourced through grant-in-aid the budget reflects the level of information, techniques and new methods, required by the function to carry out its activities. Figure 8.1 shows the level of both actual and planned expenditure in this Commission. However, the NRA is not the only organisation with a need to carry out R&D in these areas.

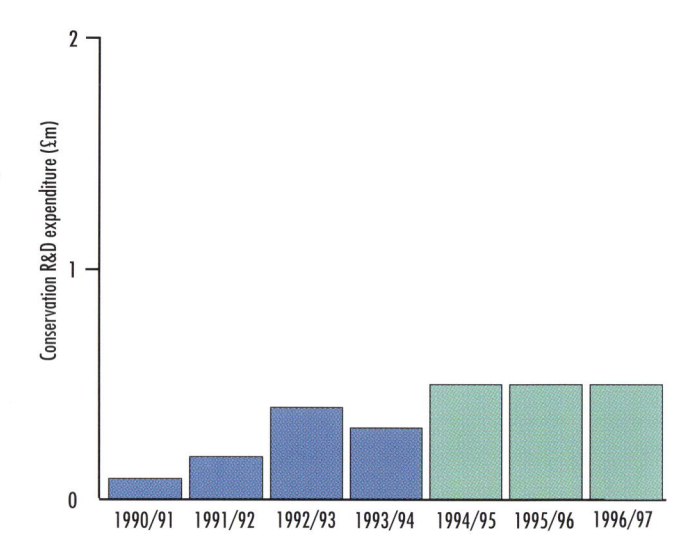

**Figure 8.1** Actual and planned expenditure on Conservation R&D up to the financial year 1996/97

# 9. CROSS-FUNCTIONAL ISSUES

This section sets out the progress made in the Commission G R&D Programme in 1993/94. Also described here are the projects for the 1994/95 financial year together with the strategic direction of the R&D required to address cross-functional issues. This area does not have a Commissioner, but is overseen on behalf of all customers by Mervyn Bramley as Head of R&D. For each project, a Director or a recognised national business group acts as sponsor.

## 9.1 Business Rationale

The NRA published its Corporate Strategy in 1994, setting out its objectives and key activities over a ten year horizon. This strategy, for the first time, sets out specific cross-functional operational issues that require addressing, and in doing so has provided an essential focus for some of the R&D that is on-going in Commission G. Of these issues, the following will drive R&D requirements in the medium-term:

- Climate change;
- Wetlands;
- River and coastal management;
- Land-use;
- Environmental impact and risk assessment; and
- Environmental economics.

Many of these issues are extremely complex and necessarily involve more detailed investigation at a functional level. However, if the NRA is to be able to address these issues through operational cross-functional management teams, integration of approaches will be essential. It is here, in particular, that R&D can be the catalyst for drawing together the functional strands of these larger issues.

## 9.2 Outputs Produced

As an organisation, dealing with day-to-day operational problems, the long-term implications of climate change may appear to be of lesser importance. However, the NRA has recognised the need to be proactive in this area and, through a study with the Institute of Hydrology, has published an R&D Report on the implications of climate change for the NRA. In implementing the findings of this report, the NRA's Board has agreed that the precautionary approach should be applied to climate change issues.

The coordination of environmental assessment issues is being continued through the implementation and testing of a draft manual of techniques. The EA manual draws upon the outputs from a number of R&D projects, and training of NRA staff will ensure that a consistent approach to this aspect of development control is fostered.

In order to summarise the recent advances in environmental economics, the NRA published R&D Report 6 in 1993. This output provided an overview of developments in environmental economics in the first four years of the NRA and was based upon a series of more detailed studies carried out for a range of other functions. The results from these projects have supported the development of an economic appraisal manual for use by NRA staff.

## 9.3 Scientific Rationale

The wide scope of projects within Commission G is drawn upon by an equally wide range of scientific expertise.

The R&D undertaken on the functional analysis of European wetland ecosystems has reviewed factors such as water level, water quality, periodic flooding, habitat preservation and species diversity. Each of these areas requires sound scientific principles to be established if effective management strategies are to be adopted.

At the other end of the spectrum, the current advances in expert systems are being used to develop a tool for NRA staff to ensure a consistent use of the NRA's financial procedures.

## 9.4 Strategic Projects

The use of airborne remote sensing for assessing the quality of coastal waters is now well established within the NRA. In order to devise maximum benefit from the information gathered, a project has been developed to abstract data on movement of coastal and estuarial sediments for Flood Defence purposes, as well as river corridor information for Conservation staff. This work supports the issues of land use and coastal management.

The NRA is also contributing to an EU LIFE project aimed at restoring areas of Norfolk Broads through the trialling of biomanipulation techniques. This work, coordinated by the Broads Authority, will involve the manipulation of aquatic habitats for the restoration of shallow eutrophic lakes.

## 9.5 Commission G Programme and Outputs

This section covers the projects undertaken within the Cross-Functional Issues Commission. The outputs that have been produced between April 1993 and May 1994 are also included.

Summary information is provided for those outputs available to the general public; a complete listing of all outputs together with their availability is provided as an insert in the back cover.

Where an output is still under consideration by the NRA, where it contains commercial in confidence information, or where it feeds into another activity such as the development of a functional manual, the current availability may be less widespread. External organisations wishing to know more about such outputs should contact the Head Office R&D Section.

The R&D Programme for the sole Topic Area in this Commission is provided in Table 9.1. This table lists the projects either on-going or proposed for starting in the financial year 1994/95. Each Regional R&D Coordinator and the R&D Section at Head Office have further details on all such projects.

A fold-out guide to the information contained in the following sections is provided on the inside front cover.

### Table 9.1 Cross-Functional Issues (Topic G1) R&D Programme (1994/95)

| Proposal/ Project No. | Title | Start/ End | Contractor | Comments |
|---|---|---|---|---|
| | **On-going projects G1** | | | |
| G01(91)8 321 | Litter in rivers and marine waters. To develop methods for determining the principal sources, pathways and sinks of litter in river corridors and on adjacent beaches. | 9/91 3/94 | Tidy Britain Group | Part-funded by EPSRC and Tidy Britain |
| F01(91)6 351 | Environmental assessments undertaken by external developers. To provide the NRA with guidance on responses to environmental assessments produced by others. Phase 2 - operational evaluation of guidance | 7/93 3/95 | WRc | |
| G01(91)1 405 | Catchment management issues, Phase 2. To develop use-related standards and supporting tools relevant to Catchment Management Planning. | 94/95 95/96 | | |
| G01(92)1 409 | Lone worker alarm - Phase 2. To provide a means by which employees working alone can summon assistance in the event of an immobilising accident or sudden illness and to study the wider potential for "out of vehicle" communications via the NRA radio system. | 4/92 3/94 94/95 95/96 | Kennedy & Donkin | |
| G01(91)5 388 | Review of specific legal issues. To undertake a study of the law and practice of the Authority in relation to fish farming, with particular emphasis upon the farming of trout, but also encompassing related difficulties arising from the farming of freshwater fish and crayfish. | 1/93 7/94 | University of Kent | |
| G01(92)6 467 | Functional Analysis of European Wetland Ecosystems. To participate in relevant wetlands management research undertaken on behalf of the European Commission (DG XII) STEP to direct EU policy in this area and to define future wetland R&D requirements for both EC and NRA. | 4/92 3/94 94/95 95/96 | University of Exeter | Part-funded by EU |
| G01(91)9 492 | Development of public perception methodology. To develop and carry out initial public perception surveys to assess public views on needs and performance. Phase 2 | 7/93 10/93 10/93 5/94 | Henley Centre for Forecasting/ BMRB | |
| G01(93)2 475 | Biomanipulation of eutrophic waters (EC LIFE). To develop a framework for the biomanipulation of aquatic habitats for the restoration of shallow eutrophic lakes and to evaluate appropriate restoration techniques. | 5/93 3/96 | In-house/ Broads Authority | Contribution to EU LIFE (DGXI) programme led by Broads Authority |
| G01(93)1 483 | Institutional aspects of European environment agencies. To evaluate the institutional framework of water resource management in a number of EC Member States to provide NRA staff with information concerning organisations working in the field of water. | 4/93 3/95 | WRc | Part-funded by EU |

| Proposal/ Project No. | Title | Start/ End | Contractor | Comments |
|---|---|---|---|---|
| G01(93)3 495 | **BSRIA Environmental Code of Practice - Phase 3** To support the national BSRIA initiative to provide an environmental code of practice for building design, operation and refurbishment. | 7/93 3/95 | BSRIA | Part-funded by BSRIA |
| G01(91)3 509 | **Development of airborne remote sensing techniques** To further develop data interpretation procedures for maximising the use of airborne remote sensing information to provide coastal sediment information for Flood Defence and a strategic overview capability for Conservation. | 2/94 7/95 | In house/ NERC | Part-funded by NERC |
| G01(93)4 518 | **West Coast Directory** To contribute to the promotion of integrated coastal zone management by providing a synthesis of existing information. | 2/94 3/96 | JNCC | |
| G01(92)5 524 | **Review of databases** To assess the availability and relevance of databases held by external organisations to NRA activities in order to improve NRA's efficiency and effectiveness. Phase 2 - review of external databases | 2/94 6/94 94/95 94/95 | In-house | |
| G01(94)7 531 | **Catchment ecosystem R&D** To examine the feasibility of undertaking a programme of research and/or development on man's impact catchment ecosystems which is both of high scientific value and of significant practical value for enabling catchments to be managed on a sustainable basis to produce detailed plans outlining the programme. | 5/94 12/94 | Queen Mary & Westfield (London) | Part-funded by NERC, EPSRC, EN and JNCC |
| | **Proposed new starts G1** | | | |
| G01(92)4 | **Land use and management change** To provide strategic information on the distribution, impact and degree of land use change on NRA's statutory duties. | 94/95 94/95 | | |
| G01(92)7 | **Expert systems for Financial Memorandum (FM)** To produce an expert system to ensure consistent and accurate adherence throughout the NRA. Phase 1 - system specification Phase 2 - system development | 94/95 94/95 94/95 95/96 | | |
| C05(91)2 | **River bank erosion protection** To carry out field trials and monitoring of selected sites to improve understanding of management practice. | 94/95 96/97 | | |
| G01(94)1 | **Risk assessment framework** To develop a broad level framework to direct NRA risk assessment studies/projects. | 94/95 94/95 | | |
| G01(94)3 | **Barrages** To assess the issues involved in barrage development to enable the NRA to respond to planning applications and in-house proposals Phase 1 - review of external best practice | 94/95 94/95 | | |
| G01(94)6 | **River management framework** To take the guiding principles outlined in the bank erosion report and apply them to a number of sites covering bank erosion, sediment transfer and various types of waterway. | 94/95 95/96 | | |

## Outputs

This section lists the principal outputs that have been produced through the Topic Programme in 1993/94.

R&D Report 6               Development of environmental economics for the NRA

R&D Note 161               Catchment management issues
                           - Use related standards

Annual R&D Review - 1994

## Summary Information

The following summaries are provided for those documents available to the general public.

**NRA Publication Ref: R&D Report 6**

**Development of Environmental Economics for the NRA**

Postle M (1993)

NRA Project No. 253

This document summaries the studies undertaken so far, as part of the R&D work on environmental economics. A brief overview if given first of what economic analysis is and how it may assist the different functional activities of the NRA. The findings of the research projects are given and an overview of the future studies is given together with a discussion of other developmental initiatives aimed at further introducing environmental economics into NRA functional activities.

Key words: Economics, environmental, charging, incentive, economic instruments.

**NRA Publication Ref: R&D Note 161**

**Catchment Management Issues: Use Related Standards**

1993

Environmental Advisory Unit

NRA Project No. 405

This report considers the need for use-related standards for a full range of uses and activities in terms of water quantity, water quality and physical features for application to Catchment Management Planning (CMP). (Standards include standard methods which make use of site-specific data.)

The report:

- identifies all current and future uses of water and the water-related environment in any catchment;

- identifies the environmental objectives and criteria required to protect the above uses;

- identifies existing standards relevant to these objectives;

- highlights areas of existing shortfall in standards;

- identifies issues for further dissemination and research.

The existing classification of uses into categories is discussed and problems with an inconsistency of approach are identified. The means of expressing the requirements of each use, i.e. as quantitative are qualitative or considered in detail.

Key words: Catchment management plan, catchment areas, criteria, environmental quality, physical features, standards, water quality and water quantity.

**NRA Publication Ref: R&D Report 12**

**Implications of Climate Change for the National Rivers Authority**

Arnell N (1993)

Institute of Hydrology

NRA Project No. 358

The possibility that global warming due to an increasing concentration of "greenhouse gases" might result in significant climate changes has been at the forefront of scientific, political and public attention since the late 1980s.

The objective of this project was to estimate the sensitivity of the NRA activities to climate change and to assist in the development of a strategy for NRA response.

The most important implications of climate change for the NRA are listed below:

- an increase in the risk of coastal flooding, and a reduction in standards of service provided by existing flood defences;

- a change in the characteristics of coastal ecosystems - higher sea levels will threaten a number of important coastal ecosystems;

- a change in the variability of river flows - this will affect many NRA interests, including water resources, water quality, fish habitat, riverine ecosystems and aesthetic quality of river corridors;

- a change in the risk of fluvial flooding; and

- a change in groundwater recharge, and hence the reliability of groundwater resources.

Key words: Climate change, flooding, recharge, river flows.

## 9.6 Commission G Budget

The funding of projects within Commission G in general reflects their multi-functional nature. Wherever possible collaborative funding from other research-commissioning organisations is sought. The level of budget allocated to Commission G is flexible in order to accommodate projects as they arise.

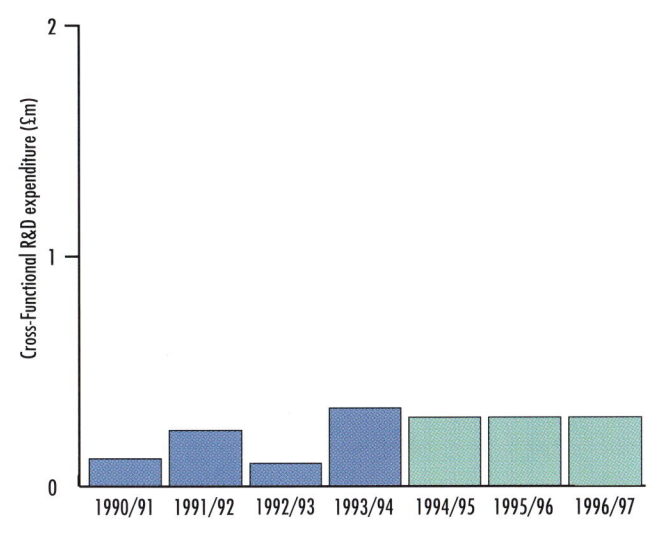

**Figure 9.1** Actual and planned expenditure on Cross-Functional Issues R&D up to the financial year 1996/97

## 9.7 Technical Services

In addition to the cross-functional R&D in Commission G, the NRA has a number of technical services which its staff can draw upon to answer urgent, specific queries. Summaries for Technical Services outputs produced in 1993/94 are given below.

**NRA Publication Ref: R&D Report 8**

**Septic Tanks and Small Sewage Treatment Works - A Guide to Current Practice and Common Problems**

Payne J A and Butler D (1993)

CIRIA

NRA Project Ref: Technical Services

Responsibility for the location, design, installation, operation and maintenance of septic tanks and small sewage treatment works is divided between planning authorities, the National Rivers Authority (NRA), building inspectors (building control officers or approved inspectors), environmental health officers and owners. However, no single authority seems able or prepared to prevent the installation of a septic tank or small sewage treatment plant in an unsuitable location, even when it is recognised that ground conditions are unsuitable and problems are likely to occur.

Several factors have been identified which currently make it difficult to reduce the occurrence of avoidable problems with new installations. One of these is the exclusion of septic tank drainage fields from Building Regulations, based on the 1947 case of Chesterton RDC v. Ralph Thompson Ltd. Whilst the relevance of this case is questioned, it has significant influence on current practice. Building inspectors are forced into accepting septic tanks in unsuitable ground conditions, as they have no powers to control effluent disposal to land.

A second factor is problems in the informal consultation process between planners and the NRA. Difficulties arise in defining when the NRA should be consulted, how it should respond, and how the response should be dealt with. Similar difficulties exist in the case of building control, since there is no formal consultation procedure between building inspectors and the NRA.

A third factor is the limitations of the British Standard which covers septic tanks and small sewage treatment works, particularly its advice on assessment of ground conditions for drainage field design.

Despite these problems there is scope for improving current practice, through better guidance, without changing legislation. It is recommended that the Department of the Environment should prepare guidance to clarify liaison procedures between planning authorities, the NRA, building control officers and approved inspectors to avoid unsatisfactory installations. Control at the planning stage is considered desirable and the use of Grampian conditions is proposed to defer planning permission until consultation procedures have been followed and satisfactory arrangements made for drainage.

Key words: Septic tanks, cesspools, package plant, groundwater protection.

**NRA Publication Ref: R&D Note 182**

**Summary Report on Environmental Developments - 13: January to March 1993**

Newman P J, Barry M K and Lewis S (1993)

WRc

NRA Project Ref: Technical Services

This report summarises environmental developments over the period January to March 1993.

Items of EC legislation introduced in this period of interest to the UK Water Industry and Regulatory Authorities include the adoption of a Council Directive relating to the harmonisation of plans to reduce waste from the titanium dioxide industry and an Amending Directive relating to the award of public contracts. Other measures introduced include an amended proposal for a Council Decision relating to the adoption of a four year programme to develop official statistics on the environment and a proposal for a Decision on reference laboratories for the monitoring of marine biotoxins.

Details of the European Commission's environmental policy priorities for 1993/94 are included in this report, policies which in many aspects mirror the position taken on the environment by Denmark, the new President of the Community. Since the last report in this series (European Developments No 12) EC Environment Ministers have met twice and have made substantial progress on several issues.

In the UK, Regulations have been introduced relating to the use and release of genetically modified organisms, the use of pentachlorophenol and the provision of information on the environment. The DoE and MAFF have issued a consultation document on the methodology for identifying sensitive areas and designating vulnerable zones. A Private Member's Bill to control noise from, amongst other things, construction and repair equipment operating in the street continues to make progress through Parliament. The Government has also indicated its intentions of expanding the scope of environmental impact assessment to include waterworks, coastal defences and trout farms.

Key words: Titanium dioxide, marine dioxins, international conferences, environmental impact assessment, pollution, genetically modified organisms, noise, information.

**NRA Publication Ref: R&D Note 205**

**Summary Report on Environmental Developments - 14: April to June 1993**

Newman P J and Barry M K (1993)

WRc

NRA Project Ref: Technical Services

This report summarises environmental developments over the period April to June 1993.

EC legislation adopted in the period, of interest to the UK Water Industry and Regulatory Authorities include, Council Regulations on an Eco-management and auditing scheme, the collection of data on existing (pre 1981) chemicals and a Resolution formally approving the Commission's Fifth Environmental Action Programme.

Commission Decisions on the analysis of mercury in fishery products and the fees payable under the proposed eco-label award were adopted. In addition, the Commission has published the first annual update of the European List of Notified Chemical Substances (ELINECS) and the Tenth Annual Report on the Quality of Bathing Water in the Community. The only new proposal for EC legislation was a proposed Directive on the health and safety of workers using chemicals, although, an amended proposal for a Directive on the Incineration of Hazardous Waste was agreed by the Council. Several important developments were forthcoming in the drafting of proposed Directives on the introduction of carbon and energy tax and the Landfill of Waste. The Commission provided some details of its plans for controlling waste tyres.

In the UK, the Royal Commission on the Environment has published a report on the incineration of waste. The proposed Contaminated Land Register has been withdrawn. Amended Regulations: the Water Supply and Sewerage Services (Customer Service Standards) (Amendment) Regulations 1993; The Control of Pollution (Exemption of Certain Discharges from Control) (Scotland) Variation Order 1993; The Control of Pollution (Registers) (Scotland) Regulations 1993; and the Control of Pollution (Discharges by Island

Councils) (Scotland) Regulations 1993, came into force during the period. Draft Regulations and Guidance notes have been published on the implementation of the Urban Waste Water Treatment Directive. Consultation Papers on proposed abstraction controls in Scotland and the designation of nitrate sensitive areas in England and Wales have also been issued for comment.

Key words: Eco-management, ELINECS, Fifth Action Programme, bathing water, waste tyres, carbon tax, incineration, nitrate sensitive areas, contaminated land, urban waste water treatment directive, network of correspondents, landfill, Germany.

**NRA Publication Ref: R&D Note 238**

**Summary Report on Environmental Developments - 15: July to September 1993**

Newman P J, Barry M K and Horth H (1993)

WRc

NRA Project Ref: Technical Services

This report summarises environmental developments in the period July to September 1993.

These include: new Directives controlling the manner in which both water companies and regulators place large supply and works contracts; a "Uniform Principle" Directive setting out the assessment criteria to be applied under the existing Plant Protection Products Directive; and the first Decisions stipulating the criteria for the award of an eco-label - those for dishwashers and washing machines.

Several important proposals for EC legislation have also been published. These include: a proposed Directive on Integrated Pollution Prevention and Control (IPC); a proposed Directive on the authorisation of biocide use; and proposed Decisions on the conclusion, on behalf of the community, of the Biodiversity Treaty, the Transboundary Water Pollution Convention and the Baltic Sea (Helsinki) Convention.

Another important development on a community level is the withdrawal of a proposed amendment to the Dangerous Substances Directive (76/464/EEC).

In the UK, and as a result of EC Legislation, new Regulations have been introduced to control the use of products containing cadmium and the supply and transport of hazardous substances.

Key words: Procurement, plant protection products, biocides, integrated pollution prevention and control, water metering, Drinking Water Directive, Bathing Water Directive, risk assessment, eco-label, Dangerous Substances Directive.

**NRA Publication Ref: R&D Note 256**

**Summary report on Environmental Developments - 16: October to December 1993**

Newman P J, Barry M K and Horth H (1993)

WRc

NRA Project Ref:. Technical Services

This report summarises environmental developments over the period October to December 1993.

The European Commission has forwarded to the Council a Communication outlining the priority areas for the distribution of EC funds via the LIFE programme in 1994. The Maastricht Treaty came into force on 1 November 1993 and will lead to significantly more EC environmental legislation being adopted by qualified majority voting. Thus in the future the response to proposed action on an EC level will increasingly require the formation of voting alliances. The decision by the Council of Ministers to locate the European Environment Agency in Copenhagen means it is now operational, although it will be well into 1994 before its exact role and powers are clarified.

In December the Council asked the Commission to review existing EC water legislation with a view to developing a more integrated policy on water management. As a part of this process a National Experts meeting was held in Brussels in December to discuss future policy on groundwater management.

In addition Environment Ministers from North Sea countries held a meeting in Copenhagen to discuss progress in achieving the goals of the Third Ministerial Conference on the North Sea.

In the UK the draft Surface Water (Fisheries Ecosystem) Regulations and the draft Water and Sewerage Services (Amendment) (Northern Ireland) Order were published for comment. Consultation papers were also published on possible amendments to the discharge and abstraction systems in England and Wales and a proposal to designate the River Dee as a water protection zone.

The Ministry of Agriculture Fisheries and Food has published a flood protection strategy for England and Wales and a report reviewing the first three years of operation of the Nitrate Sensitive Areas Scheme. The Office of Water Services (OFWAT) has published a series of reports in preparation for the 1994 Periodic Review.

Key words: LIFE, European Environment Agency, qualified majority voting, WHO Drinking Water Guidelines, draft Surface Water (Fisheries Ecosystem) (Classification) Regulations, draft Water and Sewerage Services (Amendment) (Northern Ireland) Order, Water Protection Zones, proposed Amendments to the Water Resources Act, MAFF Flood Strategy.

# APPENDIX 1

## R&D Programme for 1994/95 - Schedule of On-Going Projects and New Starts

This appendix provides a consolidated schedule of the 1994/95 R&D Programme.

Copies of the NRA's R&D Programme for 1995/96 will be available in March 1995.

| Proposal/ Project No. | Title / Objectives | Start/ End |
|---|---|---|
| | **Commission A - Water Quality** | |
| | **Business Area: Statutory Water Quality Objectives** | |
| | **Topic Area A1 - Standards and Classification Schemes** | |
| | **On-going projects A1** | |
| A04(91)7 390 | Joint Nutrient Study (JoNuS) To understand the scale, trends and processes of nutrient cycling in major east coast estuaries and coastal waters. | 11/91 1/96 |
| A04(92)2 457 | Ionisation of ammonia in estuaries To develop theoretical models which allow the concentration of toxic ammonia in estuarine waters to be calculated from a knowledge of salinity, temperature, pH and total ammonia. | 2/93 9/94 |
| A08(90)1 286 | Lakes - Monitoring and classification To develop an effective system of monitoring and classification of lakes and other standing bodies of water. Phase 2 - testing and evaluation of classification | 1/91 7/93 94/95 95/96 |
| A09(90)1 227 | Metal speciation in rivers and estuaries To determine the chemical form of selected metals in rivers and estuaries. | 7/92 5/93 6/94 3/95 |
| 037 | Development of microbial standards To define the possible effects of recreation by special interest groups in UK marine waters and to develop standards based on Phase 1. | 6/89 9/93 |
| 053 | Development of Environmental Quality Standards, Phase 2 To develop environmental quality standards (EQSs) for substances of concern to the NRA. | 7/93 5/94 |
| 051 | European pollution control philosophy To examine European pollution control strategies to assist in the development of related UK initiatives. | 6/94 3/96 |
| A(93)1 469 | General Quality Assessment Scheme To develop and test a General Quality Assessment scheme for rivers, estuaries and coastal waters to provide an objective and comprehensive classification framework for river quality surveys. | 10/93 2/95 |
| | **Proposed new starts A1** | |
| A01(94)1 | Toxicity-based criteria for assessing receiving water quality To develop and assess toxicity-based criteria in order to assess the general quality of receiving waters - Phase 1 : scoping study for marine waters, to include a review of biomarkers. | 94/95 |
| A01(94)2 | National database on recreational freshwater quality To assemble all available data on recreational freshwater quality to assist in the setting of SWQOs. | 94/95 94/95 |
| A01(94)3 | Assessment of metals in marine sediments To assess the level of metals and particle sizes within marine sediments to support the development of GQA scheme. | 94/95 94/95 |
| A08(94)5 | Faecal coliforms in shellfish and surrounding waters To assess the link between bacterial levels in shellfish flesh and surrounding waters to control pollution. | 94/95 94/95 |

| Proposal/<br>Project No. | Title<br>Objectives | Start/<br>End |
|---|---|---|

## Business Area: Monitoring
## Topic Area A2 - Strategy and Reporting

### On-going projects A2

| | | |
|---|---|---|
| A11(92)1<br>428 | **Automatic exception reporting for trends in quality**<br>To develop a semi-automatic exception reporting system, which detects trends in water quality data as soon as possible after their onset, to provide NRA staff with the means to make more effective use of all routine chemical monitoring data. | 11/92<br>10/93 |
| 015 | **Atmospheric inputs of pollutants into surface waters**<br>To determine the organic composition of atmospheric deposition in the UK and assess the contribution of airborne pollutants to the organic content of surface waters. | 10/88<br>6/94 |
| A07(92)9<br>449 | **Modelling *E. Coli* in streams**<br>To develop and apply a module for QUASAR to predict *E. Coli* in freshwater catchments. | 2/93<br>3/94 |
| A04(92)5<br>458 | **Estuarine water quality model**<br>To develop, as part of the NERC LOIS study, a portable estuarine water quality model to enable classification of waters in relation to SWQOs. | 3/92<br>4/95 |
| A(93)2<br>479 | **Improved monitoring efficiency**<br>To undertake practical catchment-based case studies to review existing monitoring activities and to develop and demonstrate new monitoring programmes and associated data analysis tools which provide an integrated approach and value for money. | 4/93<br>7/93<br><br>1/94<br>3/95 |
| A11(92)2<br>482 | **Early warning system for potential failure of river class**<br>To develop a system that would routinely inform the NRA as soon as possible of any river sampling point that was in danger of failing its class. | 7/93<br>8/94 |

### Proposed new starts A2

| | | |
|---|---|---|
| A02(94)1 | **Further testing and development of QUASAR**<br>To further develop the river quality model QUASAR to enable NRA staff to use it for specific end-user applications. | 94/95<br>95/96 |

## Business Area: Monitoring
## Topic Areas A3 - Analytical Techniques

### On-going projects A3

| | | |
|---|---|---|
| A06(91)4<br>319 | **Mothproofing agents and water quality management**<br>To develop analytical methods for and examine the fate of mothproofing pesticides in the water environment and to assess the significance of the materials when discharged to aquatic systems. | 9/91<br>9/94 |
| 062/035 | **Microbiological techniques**<br>To produce and update a manual of standard microbiological techniques for the NRA - which would include sampling analysis and quality control procedures.<br>Phase 3 - further techniques | 10/92<br>5/94<br><br>95/96<br>95/96 |
| A09(92)8<br>460 | **Use of supercritical carbon dioxide for extraction of trace organics**<br>To develop an automated method for the selective extraction of Red List organics (except organotin and dioxin compounds) at levels useful for Red List monitoring work. | 3/93<br>3/95 |
| A03(93)1<br>527 | **Farmstat pesticides**<br>To develop analytical methods based on solid phase HPLC extraction and HPLC/MS to cover the increasing range of agrochemicals used in the environment. | 6/94<br>1/96 |

### Proposed new starts A3

| | | |
|---|---|---|
| A09(92)3 | **ELISAS for the screening of herbicides and pesticides**<br>To establish a strategy for the introduction of immunoassays into the laboratories for the cost effective analysis of samples and for screening and semi-quantification of pesticides and herbicides | 94/95<br>94/95 |
| A03(94)2 | **Improved techniques for dangerous substances in saline waters**<br>To develop improved techniques to monitor dangerous substances in saline waters. | 94/95<br>94/95 |

| Proposal/ Project No. | Title / Objectives | Start/ End |
|---|---|---|
| A03(94)3 | **New techniques for pesticides**<br>To develop novel techniques for the analysis of pesticides to assist in pollution control activities. | 94/95<br>95/96 |

## Business Area: Monitoring
## Topic Area A4 - Instrumentation and Field Techniques

**On-going projects A4**

| Proposal/ Project No. | Title / Objectives | Start/ End |
|---|---|---|
| A05(91)2<br>348 | **Field detection of algal toxins**<br>To develop a field test kit for the detection to a specified level of Microcystin - LR in water. | 9/91<br>5/94 |
| A05(91)4<br>349 | **Validation of field procedures for algal toxin field test kits**<br>To validate for NRA the development and performance of the field test kit for Microcystin - LR, developed by Biocode and develop field procedures for its use by NRA staff. | 9/91<br>5/93 |
| A10(90)4<br>240 | **Bioaccumulation of Red List organic compounds**<br>To develop code(s) of practice on the use of bioaccumulation techniques for monitoring Red List trace organic substances in freshwaters and estuaries that can be used through all the NRA Regions. | 9/90<br>5/93 |
| A15(90)6<br>247 | **Broad spectrum sensors**<br>To develop and test in the laboratory and in the field a prototype instrument, incorporating biosensors, allowing a rapid assessment of toxicity of aqueous samples.<br>Phase 2 - further prototype development | 12/90<br>11/93<br><br>94/95<br>95/96 |
| A10(91)7<br>442 | **Development of new techniques for the monitoring of ammonia in water, Phase 2**<br>To develop and test in the laboratory and the field new and improved techniques for monitoring ammonia in water. | 1/93<br>9/94 |
| A10(92)1<br>427 | **Assessment of field monitors for consent monitoring**<br>To assess accuracy, reliability, applicability and cost effectiveness of available equipment to enable the NRA to monitor alternative determinands to BOD and suspended solids.<br>Phase 2 - evaluation of the best monitors | 9/92<br>2/94<br><br>94/95<br>95/96 |
| A10(92)2<br>473 | **Review of field test kits**<br>To review the use of field test kits in water quality monitoring to establish the feasibility and benefits of undertaking research to develop further test kits. | 5/93<br>9/93 |
| A04(93)2<br>507 | **Evaluation of mini metal sensors**<br>To evaluate bench-top metals analyser and to develop methods to enable environmental concentration in seawater of major saline metals to be determined *in situ* on board a survey vessel.<br>Phase 2 - production of prototype | 12/93<br>7/94<br><br>94/95<br>95/96 |
| A04(93)1<br>523 | **Moored marine water quality monitor**<br>To develop a prototype self-contained monitoring buoy in order to fulfil our statutory obligations to both monitor out to the three mile limit and to investigate the effect of polluting discharges. | 3/94<br>7/95 |
| A04(93)3<br>521 | **Feasibility study of track analysis particles**<br>To carry out a feasibility study to determine whether the CR-39 technique could be used in order to provide accurate, cost effective data for the NRA. | 2/94<br>3/94 |

**Proposed new starts A4**

| Proposal/ Project No. | Title / Objectives | Start/ End |
|---|---|---|
| A04(94)1 | **Biochemical oxygen demand predictor and hand held instrument**<br>To develop a method of indicating the level of BOD in a discharge of receiving water to enable NRA staff to make on the spot decisions for further sampling or action.<br>Phase 1 - feasibility study | 94/95<br>94/95 |
| A04(94)2 | **Oil in water - a review of existing monitors**<br>To review current equipment used to monitor the presence and levels of oil in freshwaters to provide NRA staff with accurate assessment of their performance. | 94/95<br>94/95 |
| A04(94)4 | **Inland use of airborne remote sensing**<br>To review the algorithms for interpreting data from airborne remote sensing to monitor the quality of inland waters. | 94/95<br>94/95 |

| Proposal/<br>Project No. | Title<br>Objectives | Start/<br>End |
|---|---|---|
| A04(94)5 | **Instrumentation for self-monitoring**<br>To specify and/or develop the equipment for self-monitoring and auditing of compliance to provide information for NRA staff through a business study. | 94/95<br>94/95 |

## Business Area: Monitoring
## Topic Area A5 - Biological Assessment

### On-going projects A5

| | | |
|---|---|---|
| A13(90)1<br>243 | **Testing and further development of RIVPACS, Phase 2**<br>To provide a more uniform, objective, user-friendly approach to the assessment of biological water quality for rivers within the UK. | 8/92<br>6/95 |
| A13(90)3<br>242 | **Faunal richness of headwater streams**<br>To assess the conservation value of headwater stream macroinvertebrates and their contribution to catchment macroinvertebrate richness, and determine agricultural impacts upon them and propose a conservation strategy. | 10/90<br>1/95 |
| A08(91)3<br>354 | **Effects of low level contaminants on marine and estuarine benthic communities**<br>To evaluate experimentally the effects of low levels of contaminants on marine and estuarine benthic communities. | 10/91<br>6/94 |
| A12(91)4<br>396 | **Sediment toxicity test development - insoluble substances**<br>To develop internationally standardised toxicity tests for use with sediments contaminated with sparingly water-soluble substances. | 2/92<br>3/94 |
| A12(92)3<br>420 | **Methods manual**<br>To establish standard operating procedures for ecotoxicological techniques and to produce and update a methods manual. | 9/92<br>8/95 |
| A12(92)2<br>494 | **Method development**<br>To provide suitable selection tests for the ecotoxicological assessment of effluent and receiving water quality. | 10/93<br>3/94 |
| A(93)8<br>504 | **Biological assessment methods**<br>To quantify and, where possible, control sources of variability in freshwater macroinvertebrate data for a range of river types and biological quality bands in order to increase the value of NRA data in water quality management. | 11/93<br>2/95 |

### Proposed new starts A5

| | | |
|---|---|---|
| A05(94)1 | **Applications of artificial intelligence in river quality surveys**<br>To undertake analysis of data held on the National Biological Database and to further investigate the distribution of macroinvertebrates throughout England and Wales. | 94/95<br>95/96 |
| A05(94)2 | **Biological techniques of still water quality assessment**<br>To review and examine techniques and assemblage analysis for lakes, canals and ponds; leading to classification. | 94/95<br>96/97 |
| A05(94)3 | **Alternative methods of biological classification of rivers**<br>To reassess work on macroinvertebrates and other biotic indicators in a European context. | 94/95<br>95/96 |
| A05(94)4 | **Ecotoxicological Quality Assessment procedures**<br>To develop quality assessment procedures for ecotoxicological methods and to provide accurate management information. | 94/95<br>94/95 |

## Business Area: Discharge Control and Charging
## Topic Area A6 - Consenting and Discharge Impact

### On-going projects A6

| | | |
|---|---|---|
| A01(90)4<br>305 | **Nitrification rates in rivers and estuaries**<br>To define by laboratory studies the influence on nitrification of temperature, ammonia concentration, dissolved oxygen, light, suspended solids and salinity in rivers and estuaries. | 4/91<br>1/94 |
| A04(90)6<br>339 | **Treatment process for ferruginous discharges from disused coal workings**<br>To investigate practical low cost (with respect to operation and maintenance) processes for the treatment of iron contaminated discharges from abandoned coal workings, including pilot plant trials. | 10/91<br>12/93 |

| Proposal/ Project No. | Title / Objectives | Start/ End |
|---|---|---|
| A01(92)1 432 | **Assessment and reporting of the impact of CSO discharges on receiving waters** To develop an objective assessment procedure to enable the impact of episodic events on receiving waters to be reported, within the framework of the regulatory duties of the NRA. | 12/92 3/94 |
| A01(92)3 408 | **Urban Pollution Management application methodology** To develop expertise and provide guidance for the application of the modelling tools that have been developed under the Urban Pollution Management programme. | 4/92 4/93 |
| A01(92)7 464 | **UPM Management manual** To produce a manual which will allow practitioners on both the regulator and sewerage undertaker sides of the industry to implement the Urban Pollution Management methodology to assess the impact of storm sewer overflows in rivers. | 2/93 9/94 |
| A01(92)11 436 | **Removal of coloured effluents from dyehouses** To research novel and cost effective processes for removing colour from dyehouse waste in order to minimise adverse effects on the water environment. | 1/93 10/95 |
| A01(92)9 430 | **National consent translation project** To establish a sound methodology for the neutral translation of consents for sewage treatment works' effluents into the format required by the EC Directive on Urban Waste Water Treatment. Phase 2 - standard sampler specifications | 11/92 6/94 94/95 94/95 |
| A07(92)13 456 | **Efficacy and effects of wastewater disinfection** To undertake desk, laboratory and field studies of the efficiency and environmental effects of candidate disinfection processes to support NRA policies on consenting disinfected discharges. Phase 4 - further discharge types and techniques | 2/93 6/94 94/95 95/96 |
| A(93)7 468 | **Cost benefit assessment** To develop an economic benefit methodology for evaluating environmental benefits resulting from changes in water quality stemming from improvements in effluent quality. | 7/93 3/94 94/95 94/95 |
| A06(94)6 493 | **Toxicity based consents** To further develop toxicity criteria in regulatory control, to test the protocols and procedures necessary for the application of toxicity based consents. | 10/93 3/96 |
| A07(91)3 485 | **Survival of particular viruses in seawater** To assess the extent to which particular viruses can survive in the aquatic environment after discharge in body fluids to sewage in seawater Phase 2 - practical assessments | 8/93 9/93 94/95 94/95 |
| A07(93)1 490 | **Identification of oestrogenic substances in STW effluent** To identify and quantify the component(s) present in sewage effluents that are responsible for the vitellogenic response in fish. | 9/93 3/96 |
| | **Proposed new starts A6** | |
| A06(94)1 | **Expert system for consenting discharges** To develop an expert system to assist in the setting of discharge consents: Phase 1 - definition study. Phase 2 - development of prototype | 94/95 94/95 95/96 95/96 |
| A06(94)2 | **UPM implementation project** To demonstrate the implementation of the Urban Pollution Management Manual and other products to assist NRA staff in implementing the procedures. | 94/95 95/96 |
| A06(94)3 | **Fate of detergents and associated chemicals in wastewater and rivers** To assess the sources, transport and fate of detergents and similar chemicals to enable receiving waters to be protected. | 94/95 95/96 |
| A06(94)5 | **Organic deposition and benthic effects** To produce a comparative predictive model relating the emission and settlement of suspended solids (organic carbon) to changes in the sediment dwelling microfaunal communities. | 94/95 95/96 |
| A06(94)6 | **Implications of real time control on regulation of CSOs** To undertake a scoping study to review the issues and identify the options for funding and developing knowledge of real time control on the regulation of CSOs. | 94/95 94/95 |

| Proposal/ Project No. | Title / Objectives | Start/ End |
|---|---|---|
| | **Business Area: Pollution Prevention** | |
| | **Topic Area A7 - Rural Land Use** | |
| | **On-going projects A7** | |
| A10(90)2 450 | Impact of pesticides on river ecology<br>To assess the impact of different pesticides on the structure and functioning of riverine ecosystems. Phase 1 - literature review and definition study | 3/93<br>11/93 |
| A07(91)4 424 | Occurrence of *Cryptosporidium*<br>To ascertain whether or not a link can be established in agricultural areas between livestock farm pollution incidents or farm husbandry and the occurrence of *Cryptosporidium* oocysts in samples of surface water. | 9/92<br>6/94<br><br>94/95<br>94/95 |
| A02(90)3 205 | Management of acid lakes by regulating nutrients - Phase 2<br>To obtain a viable and effective alternative methodology of habitat amelioration to liming for soft water upland lakes by addition of phosphate and monitoring changes in ecosystem and buffering capacity in order to produce a management tool. | 4/93<br>3/96 |
| A02(90)1 230 | Measures for protecting upland water quality<br>To develop management practices required for practical implementation of Forest and Water Guidelines, in particular the optimisation of buffer strip width in forest planting. | 4/90<br>6/93 |
| A02(90)4 270 | Assessing the impact of forest clear-felling on stream invertebrates<br>To determine the consequences of large scale clear-felling on upland stream communities. | 2/91<br>2/94 |
| A02(91)5 314 | Acid waters: Llyn Brianne project<br>To investigate the impact of liming treatments and land use change on streams, and to refine models for predicting deposition impacts and land use changes in the aquatic environment. | 4/91<br>3/94 |
| A02(91)4 368 | Impact of erosion of forest roads on water quality<br>To study the natural erosion processes and rates from mixed aggregate built roads in upland forests and the impacts of heavy vehicles used for timber extraction, in order to identify impact on water quality. | 11/91<br>9/94 |
| 119 | Total impact assessment of pollutants in rivers, Phase 3<br>To investigate the pollution of streams draining agricultural catchments and specifically, to develop a simple model of the movement of pesticides from the point of application to streams. | 4/92<br>5/94 |
| 122 | Effects of agricultural erosion on watercourses<br>To identify the factors affecting the yield of high sediment yielding catchments, and to produce recommendations to reduce adverse impacts of soil erosion. | 1/90<br>8/93 |
| A03(91)2 359 | Optimum application rates for low rate irrigation<br>To identify the appropriate rate, time and frequency of application of dilute farm effluent to different types of land without causing water pollution. | 11/91<br>3/94 |
| A02(92)3 465 | Impacts of fine particulate outputs associated with timber harvesting<br>To quantify the impacts of timber felling upon stream particulate loads and to investigate methods to ameliorate long and short term sediment pollution associated with forestry land use and practice. | 7/93<br>7/96 |
| A02(92)5 416 | Nitrogen module for the IH acidification model (MAGIC)<br>To develop a nitrogen process module for MAGIC and to apply the model to investigate increasing nitrate in atmospheric deposition. | 8/92<br>7/95 |
| A03(92)2 453 | Land management techniques<br>To develop management techniques for soil and nutrient conservation including the use of buffer zones and farm management plans.<br>Phase 1 - definition study<br>Phase 2 - development of models and practical techniques | 2/93<br>8/93<br><br><br>94/95<br>96/97 |
| A03(92)3 434 | Pathogens from farming practices<br>To study the incidence of pathogenic bacteria and viruses in soils receiving livestock waste and to investigate the persistence and mechanisms of transport of these organisms in the soil. | 10/92<br>10/95 |
| 001/012 | Sources, impacts and detection of farm pollution<br>To develop biological methods for the detection of organic pollution from farms and for assessing the effectiveness of remedial action. | 4/90<br>7/94 |

| Proposal/ Project No. | Title / Objectives | Start/ End |
|---|---|---|
| A02(92)10 462 | **Zinc from watercress farms** <br> To produce watercress with improved resistance to the crook root fungus *(spongospora subterranea f. sp.nasturti)* and watercress yellow spot virus in order to reduce the need to treat watercress with zinc and consequent contamination of watercourses. | 3/93 <br> 10/96 |
| A07(93)2 502 | **Impact of conifer harvesting** <br> To assess the impacts of conifer harvesting and replanting on upland stream water quality with a view to identification of ameliorative management strategies and the development of a model and guidelines for environmental impact assessment. | 1/94 <br> 12/96 |

### Proposed new starts A7

| | | |
|---|---|---|
| A07(94)1 | **Alternative farming methods - dairy** <br> To develop new systems of milk production to reduce the impact on water quality, in particular, levels of phosphate. | 94/95 <br> 98/99 |
| A07(94)2 | **Alternative farming methods - arable** <br> To develop more environmentally-friendly methods of rotational arable farming to reduce the pesticide usage. | 94/95 <br> 96/97 |
| A07(94)4 | **Pesticide disposal** <br> To develop best practice guidelines to ensure that waste pesticides are effectively disposed of. | 94/95 <br> 94/95 |
| A07(94)5 | **Guidance for assessing and controlling non-point sources of phosphorus** <br> To develop guidance for the assessment of the contribution of non-point sources to the phosphorus budget in rivers to assist in pollution prevention. | 94/95 <br> 94/95 |
| A07(94)6 | **Treatment technology for farm wastes** <br> To investigate the potential for developing new technology to treat farm wastes for disposal to land. | 94/95 <br> 95/96 |
| A07(94)7 | **Best Practice Manual** <br> To develop a manual of best practice to help NRA staff advise farmers. | 94/95 <br> 95/96 |

## Business Area: Pollution Prevention
## Topic Area A8 - Groundwater Pollution

### On-going projects A8

| | | |
|---|---|---|
| A14(91)1 381 | **Pollution potential of contaminated sites - Phase 2** <br> To assess groundwater pollution potential of contaminated sites and to relate these leaching tests to target levels for acceptability with respect to groundwater. | 94/95 <br> 95/96 |
| A07(90)1 295 | **Geochemical process modelling** <br> To identify all relevant processes, geochemical and/or biochemical, which can apply to the subsurface environment and which can be quantified for the purpose of deriving predictive transport models for operations use. | 6/90 <br> 3/94 |
| A14(91)3 380 | **Development on contaminated land - Phase 3** <br> To provide guidance to developers and regulatory bodies in the assessment, specification, supervision and achievement of effective and safe remediation of contaminated land using the most appropriate techniques. | 2/94 <br> 1/96 |
| A08(93)5 514 | **Risk assessment methodology for landfills** <br> To develop a risk assessment methodology for the assessment of landfill engineering plans. | 11/93 <br> 3/95 |
| A08(93)10 519 | **Reliability of sewers in ESAs** <br> To assess the impact of sewer leakages on groundwater quality and to identify factors which give rise to the problem and steps which can be taken to minimise the risks in order for the NRA to produce a strategy to minimise the risk of future groundwater pollution. | 2/94 <br> 11/94 |
| A08(93)11 513 | **Long-term monitoring of non-containment landfills** <br> To produce long term monitoring data for landfill gas and leachate for uncontained landfill sites on different aquifers in the UK, in order to provide the NRA and DoE with the technical background needed to develop waste management policy. | 1/94 <br> 3/95 |
| A06(94)4 528 | **Fire water and chemical spillage retention systems** <br> To develop detailed technical guidance on NRA/DoE requirements for the planning, design and construction of containment systems for the prevention of water pollution following industrial accidents. | 5/94 <br> 4/96 |

| Proposal/ Project No. | Title / Objectives | Start/ End |
|---|---|---|
| | **Proposed new starts A8** | |
| A14(92)6 | Remediation of groundwater pollution for organic solvents<br>To develop and test a decision tree to enable the NRA to respond rapidly and rationally to both short- and long-term problems concerning organic solvents. | 94/95<br>95/96 |
| A08(94)1 | Contaminated soils and shallow systems<br>To clarify the flow and transport of pollutants in soils to underpin the vulnerability classification within the GPP. | 94/95<br>95/96 |
| A08(94)2 | Effects of old landfill sites on groundwater quality<br>To assess the extent of groundwater pollution potential of old domestic waste overlying major aquifers to assist in management decisions. | 94/95<br>96/97 |
| A08(94)3 | Leachate recirculation<br>To review the extent of current knowledge on the recirculation of leachate to assess the short-term risks of water pollution. | 94/95<br>95/96 |

## Business Area: Pollution Prevention
## Topic Area A9 - Pollution Prevention

### On-going projects A9

| Proposal/ Project No. | Title / Objectives | Start/ End |
|---|---|---|
| A01(91)9<br>410 | Review of pollution emergency arrangements and remedial measures<br>To review the NRA's response and effectiveness in dealing with pollution incidents in controlled waters, and so to specify identified areas in which improvements can be made to the current emergency service across the Regions. | 6/92<br>5/93<br><br>94/95<br>95/96 |
| A08(90)4<br>271 | Production and fate of blue-green algal toxins<br>To investigate the production and fate of cyanobacterial toxins in the environment in order to elucidate effective management strategies related to cyanobacterial blooms and scums. | 12/90<br>8/94 |
| A10(90)5<br>206 | Environmental fate of persistent organic compounds<br>To investigate the occurrence of persistent organic compounds in fish flesh. | 4/90<br>3/94 |
| A01(92)10<br>425 | Control of pollution from highway drainage systems<br>To determine the impact of surface water discharges and to produce practical guidance on measures which can be taken to reduce pollution.<br>Phase 2 - new control measures | 10/92<br>12/93<br><br>94/95<br>95/96 |
| A17(90)2<br>292 | Development of a risk assessment tool for catchment control<br>To produce a pollution assessment tool, applicable on a site by site basis, to calculate the probability of that site causing unacceptable pollution to Controlled Waters. | 2/91<br>7/93<br><br>94/95<br>95/96 |
| A07(93)3<br>487 | Use of recombinant M13 bacteriophage for pollution tracing<br>To produce a method for tracing pollution sources with a readily available non-toxic organism that can be identified in the laboratory. | 8/93<br>10/93 |
| A01(92)8<br>391 | Demonstration project for industrial wastewater minimalisation<br>To identify, assess and demonstrate the benefits arising from the minimisation of industrial wastewater through systematic and strategic practice in association with other regulators.<br>Phase 2 - further demonstration areas | 4/92<br>5/94<br><br>94/95<br>96/97 |
| | **Proposed new starts A9** | |
| A02(94)2 | Distribution and dynamics of blue-green algae<br>To assess the occurrence, variability and detection of blue-green algae and their toxins as part of a pollution prevention and monitoring strategy. | 94/95<br>95/96 |
| A09(94)1 | Use of industrial by-products in road pavement foundations<br>To examine the potential for contamination of surface or groundwater as the result of using industrial by-products (including reclaimed materials) in road construction. | 94/95<br>94/95 |

| Proposal/<br>Project No. | Title<br>Objectives | Start/<br>End |
|---|---|---|

# Commission B - Water Resources

## B1 - Hydrometric Data

To develop and improve hydrometric instrumentation and related techniques, including related data storage, retrieval and processing, to ensure that methods are consistent, cost-effective and reliable throughout the NRA, and that output data can readily be utilised for management purposes.

### On-going projects B1

| | | |
|---|---|---|
| B01(92)2<br>478 | Calibration of portable electromagnetic current meters at low velocities<br>To evaluate the performance of all electromagnetic current meters currently available on the commercial market to ensure NRA staff are aware of the benefits and limitations of such meters. | 6/93<br>7/93 |
| B01(93)2<br>529 | Current metering standards<br>To improve improve hydrometric instrumentation to provide more accurate data to the NRA. | 5/94<br>3/94 |

### Proposed new starts B1

| | | |
|---|---|---|
| B01(92)4 | Enhancement of ultrasonic river flow gauges<br>To update and improve the measurement capability and accuracy of ultrasonic river flow gauges by the use of modern processes and electronic design together with improved software. | 94/95<br>95/96 |
| B01(93)1 | Performance assessment of ultrasonic and electromagnetic gauges at low velocities<br>To determine the performance of multipath ultrasonic and electromagnetic gauges at very low water velocities. | 94/95<br>95/96 |
| B01(94)1 | Evaluation of sewer flow monitors for abstraction measurement<br>To reach a decision on suitability of sewer flow monitoring equipment for monitoring abstractions and for discharges with low head, typically at fish farms and watercress beds. | 94/95<br>94/95 |
| B01(94)2 | Evaluation of acoustic doppler current profiler equipment<br>To identify the limitations of the equipment and to confirm that it operates within the manufacturer's specification to enable regional application. | 94/95<br>94/95 |
| B01(94)3 | Evaluation methodology for benefit of hydrometric networks<br>To develop the procedures for assessing (1) the requirements for hydrometric data, in an appropriate geographical area (2) the benefits derived from making such data available (3) whether or not the requirements are met by existing networks (4) what additions/reductions are necessary. | 94/95<br>95/96 |

## B2 - Flow Regimes

To improve the understanding of the inter-relationship between river flow and environmental factors with regard to both availability of water resources and impact on the aquatic environment so as to achieve a balance between water consumption and environmental protection.

### On-going projects B2

| | | |
|---|---|---|
| B02(91)3<br>316 | Effect of long term conifer afforestation and cropping in upland areas on water resources<br>To investigate the effects of forestry maturation and cropping onflow regimes and water chemistry of upland streams. | 4/91<br>3/94 |
| B2(93)3<br>491 | Design tools for low flow estimation<br>To scope the benefits and costs of a range of proposals by IH to define a programme of future collaborative research. To improve and extend the application of enhanced MICROLOWFLOWS software by selected developments. | 9/93<br>5/94<br>94/95<br>95/96 |
| B02(91)2<br>282 | Ecologically acceptable flows, Phase 2<br>To provide the framework for an objective method of evaluation of prescribed minimum flows based on the recognition of ecologically acceptable flows opposite to particular seasonal requirements of aquatic life forms. | 1/94<br>3/96 |
| B02(93)1<br>520 | Determination of minimum acceptable flows<br>To develop the concept of minimum acceptable flows (MAFs) and a policy for their application. | 2/94<br>12/95 |

| Proposal/ Project No. | Title / Objectives | Start/ End |
|---|---|---|
| B02(93)2 515 | **Extended flow records at locations in England and Wales** To synthesize monthly records at 15 locations distributed throughout England and Wales for the period 1860-1922 to be used in the assessment of the yield of water resource systems. Phase 2 | 6/94 6/94 94/95 95/96 |

### Proposed new starts B2

| Proposal/ Project No. | Title / Objectives | Start/ End |
|---|---|---|
| B02(94)1 | **Revision of SWK methodology for assessing low flows** To review and revise the new NRA methodology in terms of relevant criteria, associated data needs and scoring matrix to improve consistency of results between regions. | 94/95 94/95 |

## B3 - Water Resources Management

To develop and improve water resources assessment and management techniques to secure optimum strategies and policies for the use of water resources taking account of both catchment-wide and national interests.

### On-going projects B3

| Proposal/ Project No. | Title / Objectives | Start/ End |
|---|---|---|
| B03(92)5 484 | **Evaluating the costs and benefits of low flow alleviation - Phase 2** To undertake surveys on two specified low flow rivers to obtain data for a cost benefit analysis in order for the NRA to justify alleviating low flows. | 7/93 7/94 |
| B03(92)1 406 | **Expert systems for water resources management - Phase 2** To provide an intelligent assistant computer programincorporating an expert system and data organiser to aid abstraction licence application determination and monitoring by NRA water resources officers. | 8/93 9/94 |
| B03(93)4 414 | **Surface water yield assessment** To review and assess the suitability of existing methods for surface water yield assessment for both multiple and stand-alone sources and to recommend procedures for both sources for the NRA to adopt. | 7/92 10/93 |
| B03(93)4 505 | **Technical procedures for licence determination** To develop a methodology by which abstraction licence applications may be determined consistently and with due regard to protected rights and in-river needs. | 1/94 3/95 |

### Proposed new starts B3

| Proposal/ Project No. | Title / Objectives | Start/ End |
|---|---|---|
| B03(93)1 | **Groundwater resource reliable yield** To identify criteria and develop a methodology for estimating the yield of groundwater resources for use on a national basis to improve water resource planning. | 94/95 95/96 |
| B03(93)5 | **Metering of water abstraction - good practice manuals** To produce "Good Practice Manuals" to be used as reference documents and a basis for training, giving the best type of metering for each application plus guidance on maintenance, calibration and accuracy. | 94/95 95/96 |
| B03(94)3 | **Demand forecasting issues and methodology** To derive a forecasting methodology for Public Water Supply and industrial abstraction incorporating current initiatives within the water industry and enabling the economic level of demand management to be determined and to improve knowledge of present components of water demand. | 94/95 95/96 |
| B03(94)4 | **Incentive charges: Practical difficulties of measurement** To identify the "weights and measures" type of standards which would be required for quantity-based abstraction charges and to identify the practical problems and costs of meeting these standards by completing a field-based review for each type of abstraction including Public Water Supply, Spray Irrigation, Industry and Agriculture. | 94/95 95/96 |

| Proposal/<br>Project No. | Title<br>Objectives | Start/<br>End |
|---|---|---|

## B4 - Groundwater Protection

To understand, monitor and control the groundwater quality environment by aquifer protection policies, effective pollution control measures and by the determination of geochemical, biochemical and hydrological processes, both natural and induced, which apply in differing hydrogeological environments in time and space.

### On-going projects B4

| | | |
|---|---|---|
| B04(91)1<br>306 | **Bacterial denitrification in aquifers**<br>To investigate the controls on bacterial denitrification in the unsaturated zone of the Chalk and Sherwood Sandstone aquifers in Nitrate Sensitive Areas. | 4/90<br>5/93 |
| B04(93)1<br>499 | **National system for recharge assessment**<br>To develop an accurate, consistent and publicly defensible method of estimating mean annual groundwater recharge applicable to drift free areas of aquifer outcrop in England and Wales. | 10/93<br>10/94<br><br>94/95<br>95/96 |
| B04(92)6<br>454 | **Manual of aquifer properties**<br>To assemble and publish and Aquifer Properties Manual for England and Wales in order to underpin water resources management, particularly the groundwater protection policy. | 3/93<br>3/96 |
| B04(92)5<br>455 | **Hydrogeological characterisation of clays**<br>To assess parameters relevant to superficial clay cover, identify mission data and pursue field evaluation of techniques for measuring these data. | 2/93<br>1/96 |

### Proposed new starts B4

| | | |
|---|---|---|
| B04(93)2 | **Strategic groundwater research review**<br>To undertake a strategic review with BGS and other researchers/customers into the priority areas of applied strategic research which are of concern to the NRA and other customers, and into more strategic issues relevant to the work of BGS and other research groups. | 94/95<br>94/95<br><br>94/95<br>95/96 |

## Commission C - Flood Defence

### C1 - Fluvial Defences and Processes

To provide a better understanding of the specialised fields of engineering hydrology and hydraulics for the practising engineer to improve design and maintenance techniques across a wide range of Flood Defence Works. To ensure that River Structures are cost-effective in terms of capital investment and recurring operation and maintenance costs especially considering likely effects on the river environment due to climate change.

### On-going projects C1

| | | |
|---|---|---|
| C01(90)1<br>252 | **SERC flood channel facility, Phase 4**<br>To provide reliable procedures for assessing the hydraulic performance and stage discharge function of two-stage or compound river channels taking into account sediment movement. | 5/94<br>6/94<br><br>94/95<br>96/97 |
| C01(91)2<br>333 | **Infiltration methods for runoff control**<br>To produce a manual of good practice for the design, installation and maintenance of the range of practical infiltration techniques currently available for the effective disposal of surface water drainage. | 8/91<br>8/93 |
| C01(91)3<br>366 | **Large-scale model investigation of a two-stage channel, Phase 2**<br>To advance understanding of environmentally sensitive and cost-effective river channel design to enable appropriate flood defence standards of service to be provided. | 11/91<br>3/93 |
| C05(90)3<br>300 | **Design and operation of trash screens**<br>To establish best practice for the design and operation of trash screens, and to produce a manual of best practice.<br>Phase 3 - monitoring schemes | 10/91<br>9/93<br><br>94/95<br>96/97 |
| C05(91)1<br>384 | **Sediment and gravel bed transportation**<br>To carry out field studies and monitoring of selected sites to improve management practice. | 2/92<br>5/94 |

| Proposal/<br>Project No. | Title<br>Objectives | Start/<br>End |
|---|---|---|
| C05(91)4<br>363 | **Pumping stations - efficiency, operation and life-cycle costs**<br>To review national practices and philosophy of pump specifications, configuration, telemetry, operating rules and energy management, to identify best practice. | 12/91<br>3/94 |
| C01(91)1<br>407 | **Review of fluvial R&D related to flood defence**<br>To identify and prioritise a strategic NRA research programme in the specific area of fluvial defences and processes and to ensure that the NRA's programme interfaces with other external and internal cost programmes. | 4/92<br>3/94 |
| C01(91)1<br>394 | **Rainfall frequency studies, Phase 1b and 2**<br>To review current methods used for rainfall frequency analysis and to develop new procedures where the current methods give unsatisfactory results. To compile the new procedures to form a volume of the proposed flood estimation handbook entitled "Rainfall Frequency Estimation". | 2/94<br>5/96<br>95/96<br>95/96 |
| C01(93)1<br>508 | **Benchmarking for models**<br>To produce standard benchmarks against which to test fluvial flood defence models in order to provide the NRA with a decision-making tool on appropriateness for different applications.<br>Phase 2 - benchmarking trials | 1/94<br>7/94<br>94/95<br>95/96 |

### Proposed new starts C1

| | | |
|---|---|---|
| C01(94)1 | **Monitoring methods and techniques for fluvial defences**<br>To determine what fluvial parameters should be monitored and how to meet the needs of Flood Defence and integrate with other river management parameters. | 94/95<br>96/97 |
| C01(94)2 | **Scoping study and design notes for fluvial design manual**<br>To provide a framework, format and cross-referencing system to draw together new knowledge and techniques in a fluvial design manual. | 94/95<br>96/97 |

## C2 - River Flood Forecasting

**To support the development of systems for flood forecasting, including use of weather radar information, that will provide accurate forecasts of river stage heights with increased lead times.**

### On-going projects C2

| | | |
|---|---|---|
| C02(90)3<br>287 | **Evaluation of integrated flood forecasting models**<br>To evaluate the integrated flood forecasting systems currently being commissioned or operational in the NRA, and to provide recommendations for future development. | 1/92<br>3/95 |
| C02(91)1<br>372 | **Continuous monitoring of soil moisture for flood hydrology**<br>To install and test in the field instrumentation to provide continuous measurements of soil moisture and to assess the value of measurements from such instruments in an operational flood forecasting system. | 1/92<br>10/94 |
| C02(90)4<br>367 | **Use of vertically-pointed radar**<br>To obtain vertical radar profiles for specific sites and specific meteorological events, in order to overcome key problems associated with the existing national network of radars. | 12/91<br>5/94 |
| C02(91)2<br>357 | **Development of fully distributed models using radar rainfall data**<br>To develop the methodology to use radar data from 2 km and 5 km squares to construct distributed models of rainfall on a catchment basis. | 11/91<br>5/94 |
| C02(91)3<br>315 | **Improved adaptive calibration technique for weather radars**<br>To develop an improved adaptive calibration technique for weather radars which overcomes some existing shortcomings for hydrological forecasting. | 4/91<br>3/94 |
| C02(92)1<br>470 | **Improving the accuracy of radar rainfall data**<br>To compare accuracy of different methods of weather radar precipitation measurement and data processing techniques. | 3/93<br>11/95 |
| C02(92)2<br>448 | **Development of improved methods for snow melt forecasting**<br>To develop improved models for assessing the water content of snow packs and forecasting snow melt. | 2/93<br>2/95 |
| C02(93)2<br>510 | **Thunderstorm and deep convection warning system**<br>To develop a system for alerting hydrologists to the likely occurrence of thunderstorms and for forecasting the heavy rainfall associated with thunderstorms and deep convection through development of the GANDOLF system (Generating Advanced Nowcasts for Deployment in Operational Land-surface Flood Forecasting). | 1/94<br>3/96 |

| Proposal/ Project No. | Title / Objectives | Start/ End |
|---|---|---|

### Proposed new starts C2

| | | |
|---|---|---|
| C02(94)1 | Comparison of calibration techniques to determine best approach<br>To compare radar rainfall calibration techniques and identify best practice. | 94/95<br>95/96 |

## C3 - Catchment Appraisal and Control

To develop both existing and new approaches for assessing and influencing the impact and extent of development to prevent the creation or extension of flooding risks. To further develop the current understanding on the extent to which environmental processes may affect Flood Defence Planning.

### On-going projects C3

| | | |
|---|---|---|
| C03(91)1<br>426 | Forward planning process; Best European Practice<br>To study experiences in other EC member states and limited examples elsewhere with integrated flood plain land use and flood defence, taking into account the conservation and enhancement of the river environment. | 8/92<br>7/94 |

### Proposed new starts C3

| | | |
|---|---|---|
| C03(93)1 | Economic and environmental appraisal<br>To develop economic tools for analysing the impact of catchment control in order to link economic appraisal to environmental assessment. | 94/95<br>96/97 |
| C03(93)2 | Analysis of catchment control problems<br>To analyse the failure in planning liaison to improve the development of the control in planning of Flood Defence activities. | 94/95<br>95/96 |
| C03(94)1 | Developing new strategic instructions and guidelines<br>To provide enhanced national guidance for influencing development in flood risk areas. | 94/95<br>95/96 |
| C03(94)2 | Developers contribution to works<br>To consider the economics of development in the flood plain and determine the extent of contributions from proposed developers for NRA schemes. | 94/95<br>95/96 |

## C4 - Operational Management

To develop the framework for management of NRA flood defence maintenance, to ensure that work programmes through the NRA are consistent, prioritised adequately, justified and cost-effective, and that the interests of other functions are recognised.

### On-going projects C4

| | | |
|---|---|---|
| C04(91)2<br>341 | Asset management planning for flood defence<br>To develop a manual which identifies the variety of flood defences in existence, their strengths, failings and relative vulnerability, the management of their maintenance and, where appropriate, the methods available for rehabilitation. | 8/91<br>10/94 |
| C04(91)1<br>317 | Evaluation of alternative river maintenance strategies<br>To develop and verify a database and methodology which assesses the benefits associated with alternative river maintenance strategies in rural, mainly agricultural catchments. | 4/91<br>6/94<br><br>94/95<br>95/96 |
| C04(93)1<br>488 | Aquatic weed control operation - Phase 2<br>To produce best practice guidelines for aquatic weed control throughout the NRA in order to promote efficient and effective management practices. | 2/94<br>6/95 |
| C04(93)2<br>512 | Evaluation of Aquatic Weeds Research Unit<br>To undertake an evaluation of the past work undertaken by the AWRU for its sponsors, and to recommend how their requirements for information and advice on the management of aquatic weeds should be provided in the future. | 1/94<br>3/95 |
| C04(90)3<br>213 | Grass management operations, Phase 2<br>To assess routine mowing operations throughout the NRA; to implement further research identified in Phase 1: to carry out field trials; and to produce guidance note on best practice for riverbank grass management. | 94/95<br>95/96 |
| C04(92)3<br>435 | Economic appraisal of non-grant aided schemes<br>To develop a method for economic appraisal of flood defence works not covered by MAFF grant aid. | 1/93<br>3/94 |

| Proposal/ Project No. | Title / Objectives | Start/ End |
|---|---|---|
| C04(92)1 516 | Management of shoaling/desilting operations<br>To review guidance on the safe, economic and effective disposal and management of dredgings in order to develop best practice guidelines.<br>Phase 2 | 2/94<br>10/94<br><br>94/95<br>94/95 |

### Proposed new starts C4

| | | |
|---|---|---|
| C04(94)1 | Quality assurance for survey techniques<br>To determine what quality of survey work the NRA should specify. | 94/95<br>95/96 |

## C6 - Coastal and Tidal Defences and Processes

To ensure that cost-effective and environmentally sympathetic engineering options are adopted for coastal and estuarine flood defence through a better understanding of processes in the coastal zone. To improve the planning and execution of coastal defences by adopting coastal management techniques and developing a defence strategy to compensate for sea level rise.

### On-going projects C6

| | | |
|---|---|---|
| C06(92)1 459 | Risk assessment for sea and tidal defence structures<br>To define and indicate the use and understanding of probibalistic design of sea and tidal defence structures and to develop methods for assessing areas at risk for coastal and tidal flooding. | 2/93<br>3/96 |
| C06(90)4 279 | Use of timber in sea defence schemes<br>To review available information on different types of timber and preservative. | 1/91<br>6/93 |
| C06(92)2 446 | Beach management manual<br>To produce a practical manual incorporating current practice and research findings to direct engineers on planning, design, implementation and management of beaches and beach recharge schemes. Supporting projects for research feeding into the manual will be: | 9/93<br>9/96 |
| C06(92)13 441 | A. National database for monitoring data | 8/93<br>8/94 |
| C06(92)14 | B. Effectiveness of beach control operations | 2/93<br>9/93 |
| C06(92)15 | C. Assessment of risks of beach recharge schemes | 94/95<br>94/95 |
| C06(92)16 489 | D. Use of non-aggregate materials | 9/93<br>9/95 |
| C06(92)3 480 | Saltmarsh management for Flood Defence<br>To produce practical guidelines and further advance the understanding of saltmarshes in sea defences and consequently enhance the NRA's capability in engineering and environmental management. Packages for research feeding into the guidelines will be: | 7/93<br>6/98 |
| 444 | A. Saltmarsh management guide | 7/93<br>8/94 |
| | B. Maintenance and enhancement of saltings | 94/95<br>95/96 |
| | C. Historic changes in saltmarshes | 94/95<br>95/96 |
| | D. Experimental setback of saltings | 95/96<br>98/99 |
| | E. Estuary morphology | 94/95<br>95/96 |
| | F. Saltmarsh process | 94/95<br>97/98 |
| C06(92)8 433 | Dissemination of Anglian Sea Defence Management Study<br>To produce practical guidance notes setting out the approach, methods and conclusions in generic terms which will benefit other regions. | 11/92<br>5/94 |

| Proposal/<br>Project No. | Title<br>Objectives | Start/<br>End |
|---|---|---|
| C06(92)7<br>522 | Public safety of access to coastal structures<br>To determine the accident record of different types of coastal structures, and to identify design features which enhance public safety while retaining the engineering effectiveness. | 2/94<br>9/95 |
| | **Proposed new starts C6** | |
| C06(94)1 | Education and public awareness<br>To seek ways of explaining to the public why beach recharge and other soft defence methods are preferred. | 94/95<br>94/95 |
| C06(94)2 | Coastal management, including standards of performance for beaches<br>To identify the standard to which a beach should perform and how. | 94/95<br>96/97 |

## C8 - Response to Emergencies

To evaluate the overall standards required of the NRA in flood emergencies and to establish the best methods of response, including flood and storm tide warnings, dissemination of information and operational response procedures.

| | | |
|---|---|---|
| | **On-going projects C8** | |
| C08(91)2<br>403 | Wave input to west coast storm tide model<br>To provide wave data to be used for the study of in-shore/off-shore wave relationships for use in flood forecasting. | 5/92<br>6/93 |
| C08(91)3<br>431 | Emergency sealing of breaches<br>To investigate and develop alternative methods and materials for sealing breaches in NRA flood defences. | 4/94<br>10/94 |
| | **Proposed new starts C8** | |
| C08(94)1 | Benefit of flood warning and forecasting<br>To determine whether flood warning and forecasting systems are justified and the benefits of providing this service can be established. | 94/95<br>95/96 |

# Commission D - Fisheries

## D1 - Fisheries Resources

To undertake relevant research into the individual or population biology of fishes in order to provide basic resource information for operational management.

| | | |
|---|---|---|
| | **On-going projects D1** | |
| D01(90)1<br>256 | Eel and elver stock assessment<br>To assess elver stocks in the Severn Estuary together with elver exploitation and its implications for river stocks. | 11/90<br>10/93 |
| D01(90)3<br>244 | Development of a fisheries classification scheme<br>To enable clear analysis of stock assessment data used for operational duties and to assess performance. | 10/90<br>3/94 |
| D01(91)2<br>404 | Use of catch statistics to determine fish stock size<br>To evaluate the use of migratory salmonid, eel and freshwater fish catch statistics for the management of these stocks in England and Wales and to determine how they can best be used to estimate stock size. | 4/92<br>4/95 |
| D01(91)3<br>443 | Sea trout investigation<br>To review and evaluate current knowledge, research and stock assessment capability in relation to sea trout and to design a cost effective programme of investigations which will enable the NRA to effectively manage sea trout stocks sustainably. | 1/93<br>5/94<br>94/95<br>95/96 |
| D01(92)2<br>438 | Genetic aspects of spring run salmon<br>To review information on the genetic characteristics and time of return of spring salmon and to make recommendations on the feasibility of, and techniques for, enhancement of spring salmon. | 1/93<br>5/93 |
| D03(93)1<br>500 | Effectiveness of salmonid stocking strategy<br>To identify the most cost-effective strategies for stocking migratory salmonids in order to maximise returns of adult fish to fisheries and/or their natal river. | 11/93<br>5/94 |

| Proposal/<br>Project No. | Title<br>Objectives | Start/<br>End |
|---|---|---|
| | **Proposed new starts D1** | |
| D01(94)1 | Classification of river fisheries<br>To classify, by computer program, all waters on the Fisheries Database and to produce a computer program for the use of all Fisheries Staff in future classifications. | 94/95<br>94/95 |

## D2 - Environmental and Biological Influences

To develop an improved understanding of the effect of environmental and biological influences on fish populations so that results from monitoring programmes can be correctly interpreted and effective fisheries management carried out when natural or man-made changes occur.

| | | |
|---|---|---|
| | **On-going projects D2** | |
| 152 | River quality and fisheries status (Torridge)<br>To quantify the effects of land use change on river quality and fisheries through a case study on the River Torridge catchment. | 1/90<br>10/93 |
| D02(90)2<br>229 | Disease status of fish as an additional indicator of water quality<br>To evaluate the disease status of fish as an indicator of surface water quality, and the role of fish health factors in limiting population success. | 10/90<br>10/93 |
| D02(90)7<br>312 | Estuarine migratory behaviour of salmon and sea trout smolts<br>To investigate, using advanced telemetry techniques, and describe the behaviour of Atlantic salmon and sea trout smolts during the estuarine phase of their migration. | 1/94<br>10/95 |
| D02(91)2<br>338 | Habscore<br>To develop a stream habitat procedure that will enable prediction of salmon abundance from stream features; and which will assist the NRA to achieve related statutory objectives. | 7/91<br>3/94 |
| D02(91)1<br>429 | Coarse fish populations in large lowland rivers<br>To identify the critical factors constraining coarse fish populations in lowland rivers and to determine management strategies to develop and improve the fisheries. | 11/92<br>3/94 |
| D02(92)2<br>452 | Effects of stocked trout on the survival of wild fish populations<br>To assess the effect of stocked trout on the survival of wild fish stocks in order to produce recommendations on future trout stocking policies. | 1/92<br>9/95 |
| | **Proposed new starts D2** | |
| D02(94)1 | Setting of spawn biomass targets<br>To establish the optimum spawn biomass to maintain self-sustaining salmon and sea trout stocks to underpin the national migratory fisheries management plan. | 94/95<br>94/95 |

## D3 - Fisheries Management

To develop and implement management strategies which assist the NRA to maintain, improve, regulate and develop fisheries in ways that are adequately justified and cost-effective.

| | | |
|---|---|---|
| | **On-going projects D3** | |
| D01(90)2<br>249 | Status of rare fish<br>To determine the present status of rare fish in certain lakes of England and Wales, and to compile related information on the ecology and genetic variation of these species which is necessary to safeguard their populations. | 10/90<br>9/94 |
| D03(90)2<br>250 | Surveying by hydroacoustic techniques<br>To evaluate and develop the use of hydroacoustic techniques to quantitatively estimate the abundance and size distribution of fish in larger watercourses.<br>Phase 2 - production and use of field manual | 5/94<br>9/94 |
| D03(91)1<br>325 | Development of fish stock assessment methodologies and methods<br>To evaluate and develop efficient fishery survey strategies and methodologies and produce practical guidelines for their implementation in NRA stock assessments. | 7/91<br>5/94 |
| D03(90)1<br>370 | Design and use of fish counters and radio-telemetry<br>To evaluate the performance of automatic fish counters and their use in monitoring adult migratory salmonid stocks. | 7/91<br>9/94 |

| Proposal/ Project No. | Title<br>Objectives | Start/ End |
|---|---|---|
| D03(91)2<br>334 | Electric fishing of deep rivers<br>To develop electric fishing sampling equipment and methodology which will allow appraisal of fish stocks in large lowland rivers. | 10/91<br>3/94 |
| D03(92)3<br>440 | Survival and dispersal of stocked coarse fish<br>To evaluate the effectiveness of stocking with coarse fish, both hatchery-reared and transferred from other waters, and to determine the most effective stocking practices. | 10/92<br>4/93<br><br>94/95<br>96/97 |
| D03(93)3<br>486 | Assessing salmon stocks using a hydroacoustic counter<br>To install, operate and evaluate a hydroacoustic fish counter on the River Wye in order to produce reliable data for stock management. | 2/94<br>3/97 |
| D03(92)1<br>503 | Fish tracking developments<br>To undertake a definition study to appraise the options to develop fish tracking equipment, in particular tags and data logging systems, in order to improve the efficiency of NRA tracking studies and to obtain a greater understanding of fish biology. | 12/94<br>4/94<br><br>94/95<br>95/96 |
| D03(93)4<br>511 | Effects of retention of fish in keepnets<br>To establish the extent of physiological perturbations in fish arising from capture and confinement in a keepnet. | 1/94<br>3/94 |
| | **Proposed new starts D3** | |
| D03(92)2 | Methods for fishery rehabilitation<br>To develop new methods of environmental restoration (other than water quality related) that will enable the NRA to improve, develop and rehabilitate damaged fisheries. | 94/95<br>96/97 |
| D03(93)2 | Development of systems for designing and implementing surveys<br>To develop systems for designing and implementing fisheries survey, including evaluation of alternative strategies. | 94/95<br>95/96 |
| D03(94)1 | Smolt trapping using acoustic deflectors<br>To investigate how sound can be best used to direct the route for migration in order to increase trapping efficiency. | 94/95<br>95/96 |

# Commission E - Recreation and Navigation

## E1 - Recreation and Navigation

To develop specific directions and guidelines that will enable the NRA to promote the amenity and recreational potential of water and associated land, and to manage effectively navigation on waters and their related facilities where it is the navigation authority.

### On-going projects E1

| Proposal/ Project No. | Title<br>Objectives | Start/ End |
|---|---|---|
| E01(91)2<br>336 | Bank erosion on navigable waterways<br>To conduct field trials to identify the most appropriate technique of bank erosion protection for navigable rivers under NRA responsibility. | 7/91<br>7/93 |
| E01(91)1<br>498 | Impact of recreation on wildlife<br>To provide information to enable the NRA to reconcile its recreation and conservation duties - initial review.<br><br>Phase 2 - more detailed studies | 11/93<br>6/94<br><br>94/95<br>95/96 |
| E01(93)1<br>501 | Socio-economic review of angling<br>To assess the current status of angling in England and Wales and analyse the distribution of NRA rod licence holders to enable the NRA to plan and forecast income and research implications. | 11/93<br>4/94 |
| | **Proposed new starts E1** | |
| E01(94)2 | Demand studies - boating on NRA navigations<br>To develop an understanding of boating demand to assist the setting of charges, price structures and prioritisation of capital and revenue works. | 94/95<br>94/95 |

| Proposal/ Project No. | Title Objectives | Start/ End |
|---|---|---|
| E01(94)3 | **Use value of NRA navigations**<br>To apply "contingent valuation" methods to NRA navigations to identify visitor value to enable the NRA to better quantify benefits of the navigation function. | 94/95<br>94/95 |

# Commission F - Conservation

### F1 - Conservation Resource Appraisal and Impact Assessment

To develop effective survey, monitoring and classification methods for assessing the conservation resource of inland and coastal waters and associated lands, and to evaluate the impacts upon and clarify acceptable criteria for maintaining and enhancing the conservation status.

**On-going projects F1**

| Proposal/ Project No. | Title Objectives | Start/ End |
|---|---|---|
| F01(90)25<br>294 | **Control of invasive riparian and aquatic weeds**<br>To control the spread of invasive plants and to formulate and promote good management practices which would enhance the conservation and amenity value of areas for which the NRA is responsible. | 2/91<br>5/94 |
| F01(92)2<br>418 | **Impact assessment and acceptable conservation criteria**<br>To evaluate impacts upon and identify acceptable criteria for maintaining and enhancing the conservation status of inland and coastal waters and associated land in England and Wales.. | 94/95<br>95/96 |
| F01(92)4<br>477 | **Rehabilitation of degraded habitats**<br>To determine the effects of land use and river management practices on riverine habitats so as to more objectively define and implement an effective rehabilitation programme for degraded river corridors. | 6/93<br>3/94<br><br>94/95<br>95/96 |
| F01(93)1<br>506 | **Impact of NRA activities on archaeology**<br>To review the duties of the NRA to archaeology and the approach to be taken towards archaeological sites and discoveries, in order to enable effective execution of the NRA's statutory responsibilities in this area. | 1/94<br>3/94<br><br>94/95<br>95/96 |
| F01(93)3<br>517 | **River morphology and plant relationships**<br>To establish key relationships between physical characteristics of river types and associated aquatic plant communities by analysing data collected by NCC. | 2/94<br>5/94 |
| F01(94)2<br>526 | **Validation and calibration of River Habitat Survey data**<br>To calibrate and validate River Habitat Survey (RHS) methodology by sampling river and stream sites in England and Wales and quality-checking selected sites being surveyed by NRA staff. | 5/94<br>9/94 |
| F01(94)5<br>530 | **Working classification of River Habitat Survey**<br>To analyse RHS field survey data and generate a national river typology and habitat quality classification in order to produce a national methodology to be implemented in 1995/96. | 6/94<br>10/94 |

**Proposed new starts F1**

| Proposal/ Project No. | Title Objectives | Start/ End |
|---|---|---|
| F01(93)2 | **Conservation assessment and management strategies for ponds and other small water bodies**<br>To develop a cost-effective, predictive means of assessing the conservation status of ponds and to identify management techniques which optimise aquatic and associated terrestrial habitats. | 94/95<br>95/96 |
| F01(94)1 | **River channel typology for catchment and river management**<br>To develop a standard method of surveying channel cross-sections with respect to form an associated vegetation/habitat structure, in a fashion amenable to easy data management and computerisation. | 94/95<br>94/95 |
| F01(94)4 | **Investigation of the distribution and impact of a fungal disease of alders**<br>To establish current and future distribution of the disease, identify potential impacts on the river corridor and recommend management options for alleviating impacts. | 94/95<br>94/95 |

| Proposal/ Project No. | Title Objectives | Start/ End |
|---|---|---|
| | **F2 - Conservation Management** | |
| | To develop management strategies and techniques which are consistent and effective from the Conservation viewpoint, compatible with the NRA's other duties where possible, and justifiable in terms of costs and the NRA's statutory responsibilities. | |
| | **On-going projects F2** | |
| F01(91)11 400 | Appraisal of conservation enhancement of flood defence works<br>To develop a method for post-project appraisal of habitat conservation and enhancement works and to assess the value of such works in relation to natural recovery from NRA operational schemes. | 3/92 3/95 |
| F01(91)10 352 | Aquatic flora database<br>To develop and update a computerised database of submerged and floating aquatic vascular plants in order to provide up to date information on their occurence and distribution. | 10/91 3/95 |
| F01(91)1 378 | Conservation of freshwater crayfish<br>To assess the impact of introductions of non-native crayfish and outbreaks of crayfish plague on freshwater ecosystems and to formulate a strategy for the conservation of the native species (*Austropotamobius pallipes*). | 1/92 12/94 |
| F01(91)2 383 | Wetland creation/river corridor enhancement<br>To assess the conservation value of 'off-river' wetland habitats created during a river corridor enhancement programme and investigate experimental management techniques for increasing the native conservation value of created habitats. | 1/92 1/95 |
| A08(91)5 346 | Physical environment for river invertebrate communities<br>To develop a unified method for the ecological assessment of "functional habitat" analysis for use by Flood Defence engineers and Water Quality and Conservation scientists. | 4/91 3/94 |
| F02(92)3 461 | Species management in aquatic habitats<br>To identify effective management strategies for protecting and enhancing the status of rare or dispersed species and minimising the impact of nuisance species, associated with inland and coastal waters in England and Wales. | 3/93 9/93 94/95 95/96 |
| A02(91)1 313 | Classification of catchment sources in relation to the ecological effects of wetland liming<br>To refine the understanding of the potential biological effects of catchment liming on important conservation resources and produce a classification system for upland wetlands of different conservation value and susceptibility. | 7/91 5/94 |
| F02(92)1 472 | Conservation in coastal areas<br>To clearly define and interpret NRA legal responsibilities in the coastal zone and relate these to the legal responsibilities of other relevant organisations and to NRA's own conservation duties. | 5/93 8/93 94/95 95/96 |
| F02(92)2 474 | Wetland conservation<br>To identify the wetland resource and develop effective management strategies to define the NRA's role in the protection, rehabilitation and creation of wetlands in England and Wales. | 5/93 3/94 94/95 95/96 |
| F02(93)2 525 | Survey of habitats for invertebrates on exposed riverside sediments<br>To review information obtained from river corridor habitat and other NRA surveys on habitat features in order to assess their likely importance. | 5/94 2/95 |
| | **No proposed new starts F2** | |

| Proposal/<br>Project No. | Title<br>Objectives | Start/<br>End |
|---|---|---|

## Commission G - Cross-Functional Issues

### G1 - Cross-Functional Issues

The Cross-functional Commission covers major issues relating to more than one of the NRA's primary functions. Projects undertaken in this Commission must have a nationally-acknowledged customer or sponsor and be accepted by the Commissioners whose functions are charged.

**On-going projects G1**

| | | |
|---|---|---|
| G01(91)8<br>321 | **Litter in rivers and marine waters**<br>To develop methods for determining the principal sources, pathways and sinks of litter in river corridors and on adjacent beaches. | 9/91<br>3/94 |
| F01(91)6<br>351 | **Environmental assessments undertaken by external developers**<br>To provide the NRA with guidance on responses to environmental assessments produced by others.<br>Phase 2 - operational evaluation of guidance | 7/93<br>3/95 |
| G01(91)1<br>405 | **Catchment management issues, Phase 2**<br>To develop use-related standards and supporting tools relevant to Catchment Management Planning. | 94/95<br>95/96 |
| G01(92)1<br>409 | **Lone worker alarm - Phase 2**<br>To provide a means by which employees working alone can summon assistance in the event of an immobilising accident or sudden illness and to study the wider potential for "out of vehicle" communications via the NRA radio system. | 4/92<br>3/94<br><br>94/95<br>95/96 |
| G01(91)5<br>388 | **Review of specific legal issues**<br>To undertake a study of the law and practice of the Authority in relation to fish farming, with particular emphasis upon the farming of trout, but also encompassing related difficulties arising from the farming of freshwater fish and crayfish. | 1/93<br>7/94 |
| G01(92)6<br>467 | **Functional Analysis of European Wetland Ecosystems**<br>To participate in relevant wetlands management research undertaken on behalf of the European Commission (DG XII) STEP to direct EU policy in this area and to define future wetland R&D requirements for both EC and NRA. | 4/92<br>3/94<br><br>94/95<br>95/96 |
| G01(91)9<br>492 | **Development of public perception methodology**<br>To develop and carry out initial public perception surveys to assess public views on needs and performance.<br><br>Phase 2 | 7/93<br>10/93<br><br>10/93<br>5/94 |
| G01(93)2<br>475 | **Biomanipulation of eutrophic waters (EC LIFE)**<br>To develop a framework for the biomanipulation of aquatic habitats for the restoration of shallow eutrophic lakes and to evaluate appropriate restoration techniques. | 5/93<br>3/96 |
| G01(93)1<br>483 | **Institutional aspects of European environment agencies**<br>To evaluate the institutional framework of water resource management in a number of EC Member States to provide NRA staff with information concerning organisations working in the field of water. | 4/93<br>3/95 |
| G01(93)3<br>495 | **BSRIA Environmental Code of Practice - Phase 3**<br>To support the national BSRIA initiative to provide an environmental code of practice for building design, operation and refurbishment. | 7/93<br>3/95 |
| G01(91)3<br>509 | **Development of airborne remote sensing techniques**<br>To further develop data interpretation procedures for maximising the use of airborne remote sensing information to provide coastal sediment information for Flood Defence and a strategic overview capability for Conservation. | 2/94<br>7/95 |
| G01(93)4<br>518 | **West Coast Directory**<br>To contribute to the promotion of integrated coastal zone management by providing a synthesis of existing information. | 2/94<br>3/96 |
| G01(92)5<br>524 | **Review of databases**<br>To assess the availability and relevance of databases held by external organisations to NRA activities in order to improve NRA's efficiency and effectiveness.<br>Phase 2 - review of external databases | 2/94<br>6/94<br><br>94/95<br>94/95 |

| Proposal/<br>Project No. | Title<br>Objectives | Start/<br>End |
|---|---|---|
| G01(94)7<br>531 | **Catchment ecosystem R&D**<br>To examine the feasibility of undertaking a programme of research and/or development on man's impact catchment ecosystems which is both of high scientific value and of significant practical value for enabling catchments to be managed on a sustainable basis to produce detailed plans outlining the programme. | 5/94<br>12/94 |
| | **Proposed new starts G1** | |
| G01(92)4 | **Land use and management change**<br>To provide strategic information on the distribution, impact and degree of land use change on NRA's statutory duties. | 94/95<br>94/95 |
| G01(92)7 | **Expert systems for Financial Memorandum (FM)**<br>To produce an expert system to ensure consistent and accurate adherence throughout the NRA.<br>Phase 1 - system specification<br>Phase 2 - system development | 94/95<br>94/95<br><br>94/95<br>95/96 |
| C05(91)2 | **River bank erosion protection**<br>To carry out field trials and monitoring of selected sites to improve understanding of management practice. | 94/95<br>96/97 |
| G01(94)1 | **Risk assessment framework**<br>To develop a broad level framework to direct NRA risk assessment studies/projects. | 94/95<br>94/95 |
| G01(94)3 | **Barrages**<br>To assess the issues involved in barrage development to enable the NRA to respond to planning applications and in-house proposals<br>Phase 1 - review of external best practice | 94/95<br>94/95 |
| G01(94)6 | **River management framework**<br>To take the guiding principles outlined in the bank erosion report and apply them to a number of sites covering bank erosion, sediment transfer and various types of waterway. | 94/95<br>95/96 |

# APPENDIX 2

## R&D Personnel and Management Information

This appendix provides details of those staff involved in the NRA's R&D Programme. R&D is an integral part of the NRA's activities and consequently, the participation of Commissioners, Topic Leaders and Project Leaders has been recognised as being central to the successful delivery of the required outputs.

## Table A2.1 Composition of Research & Development Committee[1]

### Head Office

| | |
|---|---|
| Mervyn Bramley | Head of R&D |

### Regional R&D Contact Points

| | |
|---|---|
| Mick Pearson | Regional Water Quality Manager, Anglian |
| Malcolm Colley | Technical Manager, Northumbria & Yorkshire |
| Chris Newton | Environmental Quality Manager, North West |
| Andrew Skinner | Regional Technical Manager, Severn-Trent |
| Jim Wharfe | Regional Scientist, Southern |
| Richard Symons | Planning Manager, South Western |
| Roger Sweeting | Regional Scientist, Thames |
| Charlie Pattinson | Regional Technical Manager, Welsh |

### Functional Committee Representatives (Commissioners)

| | |
|---|---|
| John Seager | Head of Environmental Quality, Head Office |
| Richard Streeter | Water Resources Officer, Head Office |
| Lindsay Pickles | Flood Defence Officer, Head Office |
| Chris Mills | Area Fisheries, Conservation and Recreation Manager, North West |
| Paul Raven | Conservation Officer, Head Office |

### R&D Officers Attending

| | |
|---|---|
| Gareth Llewellyn | R&D Planning Officer, Head Office |
| John Dalton | R&D Programme Officer, Head Office |

### Corresponding Members

| | |
|---|---|
| Clive Swinnerton | Director - Water Management and Science, Head Office |
| Howard Pearce | Head of Corporate Planning, Head Office |

[1] As at October 1994

## Table A2.2 Commissioners and Topic Leaders

| Topic Area | | Topic Leader | Region |
|---|---|---|---|
| **Commission A - Water Quality;** | R&D Commissioner John Seager (Head of Environmental Quality - Head Office) | | |
| A1 | Standards and classification schemes | Mark Everard | Head Office |
| A2 | Monitoring - strategy and reporting | Jacqui Gough | Head Office |
| A3 | Analytical techniques | Dave Britnell | Thames |
| A4 | Instrumentation and field techniques | Paul Williams | South Western (Twerton) |
| A5 | Biological assessment | Roger Sweeting | Thames |
| A6 | Consenting and discharge impact | Gerard Morris | Northumbria & Yorkshire (Olympia House) |
| A7 | Rural land use | Bob Huggins | South Western (Blandford) |
| A8 | Groundwater pollution | Bob Harris | Severn-Trent |
| A9 | Pollution prevention - general | Phil Chatfield | Thames |
| **Commission B - Water Resources;** | R&D Commissioner Richard Streeter (Water Resources Officer - Head Office) | | |
| B1 | Hydrometric data | Geoff Burrow[1] | Southern |
| B2 | Flow regimes | Mike Owen | Thames |
| B3 | Water resources management | Cameron Thomas | Anglian |
| B4 | Groundwater protection | Mike Eggboro | North West |
| **Commission C - Flood Defence;** | R&D Commissioner Lindsay Pickles (Flood Defence Officer - Head Office) | | |
| C1 | Fluvial defences and processes | David Wilkes | Thames (Barrier) |
| C2 | River flood forecasting | Bob Hatton | South Western |
| C3 | Catchment appraisal and control | John Gardiner[2] | Thames |
| C4 | Operational management | Gary Lane | Southern |
| C6 | Coastal and tidal defences/processes | Robert Runcie | Anglian |
| C8 | Response to emergencies | Senaka Jayasinghe | Thames |
| **Commission D - Fisheries;** | R&D Commissioner Chris Mills (Area Fisheries, Conservation & Recreation Manager, North West) | | |
| D1 | Fisheries resources | Alan Winstone | Welsh |
| D2 | Environmental/biological influences | Nigel Milner | Welsh (Caernarfon) |
| D3 | Fisheries management | Steve Bailey | Northumbria & Yorkshire |
| **Commissions E & F:** | R&D Commissioner Dr Paul Raven (Conservation Officer - Head Office) | | |
| **Commission E - Recreation and Navigation** | | | |
| E1 | Recreation and Navigation | Craig McGarvey | Head Office |
| **Commission F - Conservation** | | | |
| F1 | Conservation resource appraisal and impact assessment | Peter Barham | Anglian |
| F2 | Conservation management | Richard Howell | Welsh |
| **Commission G - Cross Functional - Co-ordinated through R&D Section - Head Office** | | | |

[1] Heads new Topic B5 - Demand Management - from October 1994. New Topic Leader John Adams, North West.
[2] Transferred to Commission G October 1994 - joint Topic Leaders John Gardiner and Hugh Howes.

## Table A2.3 Regional R&D Coordinators

| Coordinator | Region |
|---|---|
| Geoff Brighty[1] | Anglian |
| John Dalton[2] | Head Office |
| Mike Briers | Northumbria & Yorkshire (Leeds) |
| Jim McEvoy | North West |
| Bogus Zaba | Severn-Trent |
| Tessa Crawshaw[3] | Southern |
| Alan Burrows | South Western (Exeter) |
| Maxine Forshaw | Thames |
| Henry Brown | Welsh |

[1] Seconded to other assignment until January 1995; John Cocker acting R&D Coordinator.
[2] R&D Programme Officer acting as R&D Coordinator for Head Office.
[3] Left the NRA October 1994; temporary replacement acting as R&D Coordinator..

# Table A2.4 R&D Project Leaders for the 1994/95 Programme

### ANGLIAN

C Ashcroft, A Barnden, P Barham, G Brighty, A Bullivant, D Dunn, A Ferguson, I Forbes, P Grange, M Grout, B Harbott, I Hirst, S Jeavons, D Leggett, M Pearson, G Phillips, R Runcie, D Tester, C Thomas, R Vallentine, P Waldron

### HEAD OFFICE

P Bird, J Dalton, M Everard, J Gough, M Griffiths, C Hager, G Llewellyn, T Long, C Martin, G Mawle, C McGarvey, L Pickles, M Postle, P Raven, M Roddan, J Sherriff

### NORTHUMBRIA & YORKSHIRE

J Aldrick, R Armitage, S Axford, S Bailey, A Barraclough, G Bird, L Bird, M Briers, E Chalk, R Cresswell, J Cross, M Daniel, G Edwards, R Freestone, B Hemsley-Flint, J Hogger, J Housham, G Morris, J Pygott, K Schofield, C Turner, C Urquhart, D Wilkes

### NORTH WEST

J Adams, M Aprahamian, P Birchall, M Eggboro, J Greaves, P Jones, P Kerr, D Major, J Mawdesley, J McEvoy, R Moore, E Mycock, G Noonan, I Pearce, A Taylor, M Walsingham, P Williams, A Withers

### SEVERN-TRENT

V Brown, A Churchwood, P Crockett, J Everard, S Fletcher, R Goodhew, R Harris, R Harvey, P Hickley, S Howard, G Lane, P Lidgett, D Martin, D Pettifer, S Powers, P Robinson, P Stewart, C Thomas, J Waters, P Whalley, S Wills, D Woodcock

### SOUTHERN

K Bedford, G Burrows, T Crawshaw, R Dines, J Frake, N Haylor, D Lowthion, J Morgan, O Pollard, P Shaw, S Taylor, C Terry, S Turner, J Wharfe

### SOUTH WESTERN

L Aucott, S Bray, B Brown, J Driver, R Grantham, R Hatton, N Holden, R Horrocks, R Huggins, L Jenkins, P Monk, J Murray-Bligh, A Newman, T Newman, N Pallister, J Proctor, R Robinson, R Smith, N Stevens, C Tubb, S Wood

### THAMES

A Brookes, N Bray, P Burrows, A Butterworth, A Driver, J Eastwood, J Gill, D Greenaway, B Greenfield, C Haggett, J Harper, S Killeen, D Leeming, P Logan, B McGlashan, G Merrick, M Owen, R Pethick, V Robinson, D Rylands, J Steele, R Sweeting, J Thomas, D Tinsley, D Van Beeston, M Whiting, D Willis, B Winter, D Vickers

### WELSH

D Anderson, I Barker, C Bolton, I Davidson, G Davies, R Fisher, P Gough, F Jones, S Halfacree, R Merriman, M Mills, R Milne, N Milner, W Purvis, A Rees, N Reynolds, H Roberts, E Roblin, I Thomas, N Weatherley, A Winstone